A STUDY OF THE EFFECT OF PROLACTIN 125 I – hPRL ON ITS RECEPTORS IN BENIGN AND MALIGNANT BREAST TUMORS

SAMI AL- MUDHAFFAR

IBTEIIAL KHUDIER AL- ZAl'ADEE

CHAPTER ONE
LITERATURE SURVEY

Preface

The breast is the most common site of benign and malignant Tumours in women 0 Among their life ,about 50% and 10% of women are suffering from benign and malignant tumours respectively (1) .

Benign tumours are associated with a high probability of the development of breast cancer in women who have a history of benign tumours in their breast (2) .

Breast cancer is the common cause of death in middle- aged women And affects nearly a million women worldwide each year 0 About 35,000 deaths were occurred in women in the united states in 1983 , while there were about 41,000 deaths from this disease in 1987 (1,3)0

It appears that a large number of women are suffering from this disease, However, the various studies are essential for the improvement of diagnosis and treatment of breast cancer .

1.1: Breast Cancer :-

It is frequent in women of all ages over the last 30 years, the probability of developing the disease increases throughout the life and the mean age of women suffering from breast cancer is 60- 61. The disease is more common in whites than nonwhites . The incidence of the disease among nonwhits is mostly blacks however, is increasing specially in younger women (4) .

1.1.1:Etiology of Breast Cancer :-

Like other types of cancer , the causes of breast cancer are not well understood but the most important of these causes are (5} :
1. Genetic predisposition
2. Immunologic deficiency
3. Viruses and other carcinogens .

Several other factors are also associated with increased risk of developing breast cancer in women , such as (6,7) :
1. The age (older than 40 years) .
2. Family history : first-degree relatives .
3. Obstetric history : late parity (older than 35).
4. Previous cancer in one breast .
5. Others organ cancer , such as endometrium and ovary.
6. Fibrocyctic changes- proliferative type .
7. Excess exposure of the breast to radiation .

1.1.2:Symptoms and Signs of Breast Cancer (8) :-

The presenting complaint in about 70% of patients with breast cancer is a lump (usually painless) in the breast . About 90% of breast masses are discovered by patients herself less frequent symptoms are breast pain , nipple discharge , erosion , retraction, enlargement, itching of the nipple and redness , an axillary mass , swelling of the arm , or bone pain (from metastasis) may be the first symptom .
Breast cancer is the common cause of death in middle- aged women

The relative frequency of carcinoma in various sites in the breast is shown in fig. 1-1. Almost half of cancer of the breast being in the upper outer quadrant. A high percentage is present also in the central portion and it is due to the inclusion of cancers that spread to the subareolar reagion from neighboring quadrants . Cancer is slightly more common in the left breast than in the right .

1.1.3:Early Detection and Diagnosis of Breast Cancer :-

Early detection of breast cancer before it has spread to axillary lumph nodes , increases the chance of survival . About 84% of women whom breast cancer were detected early survive at least 5 years (9) . Symptoms and signs of breast cancer are used for the diagnosis of the disease . The confirmed diagnosis depends ultimately upon examination of the tissue removed by biopsy , the method may be done either by open biopsy under local anesthesia or by a fine needle aspiration (10) . Biopsy examination is important for diagnosis and staging of the tumours . Clinical and histologic staging of tumours are of prognostic significance and used for the designing of treatment plan (11) . Staging is also important for the assessment of survival period of breast cancer patients (12).

Several investigations other than biopsy are helpful in the diagnosis of breast cancer , including (13) :-

1. Mammography
2. Chest x- ray
3. Bone and liver scans utilizing technetium Tc 99 m tabelled phosphates or phosphonates

1.1.4: Pathologic Types of Breast Cancer:-

Four types of mammary carcinoma have been identified on the basis of cellular differentiation and invasiveness they are (14).

1. Type I lesions as exemplified by noninvasive intraductal papillary and lobular carcinoma rarely metastasize (about 1% of cases have positive axillary lumph nodes) .

2. Type II cancers as exemplified by invasive but relanvely well differentiated tumours metastasize more frequently (about 34% positive lumph nodes)

3. Type III and iv lesions are generally less well differentiated and have a greater tendency to metastasize (55- 60% positive lumph nodes) The relative frequency of the various pathologic types (type I 5% type II 15% type III 65% type Iv 15%) is that about 80% of breast tumours are of the invasive metastasizing (14)

I.I.5: Treatment of Breast Cancer:-

Treatment may be curative or palliative Curative treatment is advised for stage I and II disease treatment can be palliative for patients in stage III and iv and for previously treated patients who develop distant

Metastasis or unresectable local recurrence (15) The extent of disease and its aggressiveness are the major factors influenced the primary therapy (16) .

Several approaches have been suggested to the plan of treatment of breast cancer (17) .

1.1.S.1:Surgery :-

Patients are treated by segmentectomy (lumpectomy) when the tumour is less than 3 em wide and by total mastectomy when the tumour is bigger or centrally located surgery is followed by radiation therapy on the chest wall after total mastectomy or on the remaining breast tissue after segmentectomy (17).

1.1.5.2:Radiotherapy :-

Radiotherapy used after mastectomy or segmentectomy particularly for tumours in stage I and II in this way the tumours is controlled locally for 97% of stage Iwomen and 87% of stage II women. The survival rate is 93% for stage Iand 84% for stage II (18).

1.1.5.1 3:Hormonal Therapy :-

Hormonal therapy is the treatment of choice in hormone related breast cancer . Tamoxifen is the most widely used hormonal treatment in breast cancer (19) . Tamoxifen is an antagonist to the estrogen receptors . It is effective in both premenopausal and postmenopausl women , but its effect is less in younger women 920) . Tamoxifen action depends principally on the presence of estrogen receptors in breast cancer tumours and could be used for the treatment of both early and metastatic breast cancer 912).

1.1.5.4:Chemotherapy :-

Chemotherapy is used for the treatment of advanced or metastatic breast cancer (22) .Several cytotoxic agents have been used either singly or in combination for the treatment of advanced breast cancer . The active agents are antimetabolites , 5-fluorouracil and methotrexate , alkylating agents such as cyciophosphamide and melphalan and cytotoxic antibiotics such as mitomycin c (23) .

Chemotherapy may be used as adjuvant therapy particularly in premenopausal patients . The rational of such therapy is based on the premise that minimal residual disease after primary resection of a carcinoma may be eliminated by early systemic therapy 924)

I.I.S.S:Biological and Gene Therapy :-

As yet biological and gene therapy have made no impact on the clinical management of breast cancer . Techniques such as immunological targeting growth factor manipulation and gene therapy may well hold the future for breast cancer (25).

1.2:Beniqn Breast Tumours :-

Among their , about 50% of women may suffer from the symptoms of benign tumours (26) . Two thirds of the tumours found during a womens reproductive years are benign and represent fibrocystic changes , fibroadenoma and papillomas (27) .

1.2.1:Fibrocystic Changes:(Mammary Dysplasia)

It is common in women 30 – 50 years of age but rate in postmenopausal women and may be the most frequent lesion of the breast . The finding of fibrocystic disease include cysts (gross and microscopic) . adenosis fibrosis and ductal epithelial hyperplasia (28).

Mammary dysplasia may produce an asymptomatic lump in the breast but pain or tenderness often calls attention to the mass . Fluctuation in size and rapid appearance disappearance of the breast tumour are common in cystic disease (29) .

1.2.2:Fibroadenoma :-

Another common benign lesion is the fibroadenoma , which appears predominantly in young women and adolescents (30) . It is initially seen as a firm, mobile mass and may be very large particularly in adolescents .

Fibroadenoma are multiple and bilateral in about 14% to 25% of patients (31).

1.2.3:Papilloma :-

Papilloma manifests as a serous , serosanguineous , or watery type of nipple discharge . in the absence of a mass, the most common cause of bloody nipple discharge is an intraductal papilloma (32) . papilloma lesions may be as large as 4 or 5 em , and on gross examination , papillomas are tan or pink viable tumours (33) . Available evidence suggests that these lesions rarely undergo malignant transformation (34)

1.2.4:Treatment of Benign Breast Tumours :-

Diagnosis of benign breast tumours are confirmed by biopsy treatment of benign breast tumours is essential because of the following reasonse (36).
1. Control of clinical symptoms .
2. Normalize dense and nodular breasts
3. Reduce the risk of cancer

Three methods have been suggested to treate benign breast tumours they are (36).
1. Local exision .
2. Diet and vitamin therapy
3. Hormonal manipulation

Local exision may be used for the treatment of fibrocystic changes and fibroadenoma (37) . Dimethyl xanthines (caffeine and theophylline) and nicotin have been suggested to stimulate the formation of fibrocystic changes in the breast consequently patients must stope consuming coffee tea , cola and etc. (38) , vitamine E have been advocated to be helpful in relieving symptoms and causing regression of fibrocystic change of the breast (39) .

Hormonal treatment is also used in the manipulation of benign breast tumours for instance danazol an impeded androgen has been used for the treatment of benign breast tumours in addition tamoxifen is used also for the same purpose (40) .The last and most recent medical treatment has advocated the use of bromocriptine (41).

2.3 biochemical aspect of breast tumors

In normal cells *tlw:* biochemtcal events art:; well controlled and the tnvolved biomolecules ara in their normal levels In tumour cells there are less regulation, accordingly several types of' biomolecules may *vary* in concentration and nature (42) .

Diff8rences in enzyme activity , hormone levels and their receptors between normal and cancerous tissues have been sought to explain the phenomenon of unc ntrolled growth (malignant and benig (43). Biochemical studies of malignant and benign tumours may support the following (44) :

1. Suggestion of the presence of cancer and identification the involved organ.

2. The extent and stage of cancer .

3. Evaluation the progression or regression of the disease by diniciane.

4. Prediction whether therapeutic agent will be effective .

Several enzymes and their isoenzymes are studied in breast cancer patients, they include : ribonuclease (RNase) (45), sialyltransferase (46), and galactosyltransferase(47), fucosyltransferase(48)and lactate dehydrogenase (49). Most of these enzymes revealed variations of clinical significance.

Hormones also vestigated in breast cancer patients , they include :LH , F l·t•.estr io_l, progesterone.,hCG and PL (50,51). These hormones showed · marked variations and may used in the clinical evaluation of the disease.The most important trait of biochemical studies in breast cancer is the estrogen

1.4: Prolactin end BreiJst Tumours:-

1.4.1: General Consideration =··

Prolactin (PRL) is a protein hormone with a molet.:ular "',

about 23.000 dalton It is contained a single polypeptidr. :::'o•" ... cornpnsed 199 ammu a;:td residues Prolactin secreted by lactorr cJplu:s vv··· clre acidophilic cell:> 111 tile anterior pituitary (54) The structure ofhiJrli..-J · pro101r;tui 1s s1milar to tllat of grnwth hormon (GH),and for a long tnne tt v' • · belleVf:Ci tllett the pit;Hlarv or many antmals, d1d not contain prolactin Hovve.,. lluman proia•:un has n;Jv.; bee:r; ISOidt(:d and its structure determined ᵣe\ air· many structural stml!cHities with 11uman growth llormone \7-5.\

1.4.2:Assay of Prolactin :-

Numours metr1ods ha-.,e been Sltggested for the detern ·natio: '1+ ᵣᵣᵣᵣ'·· in tissues and cirr.ulation . they are (56) ·

1 . Bioassay ·

2. Radioimmunoassay (RIA)

3. ImmunoradiornetnG assay (IRMJI.J

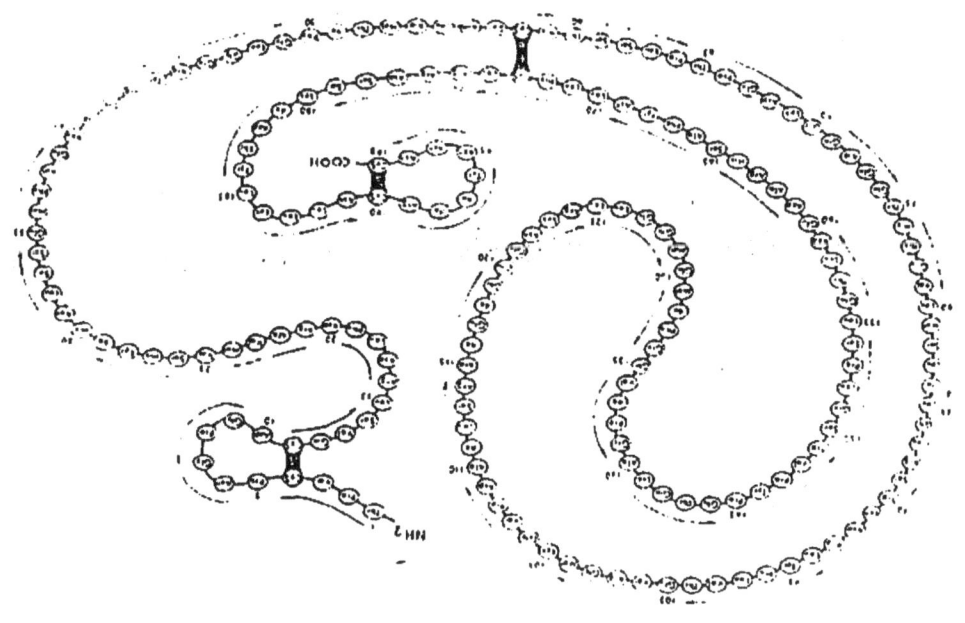

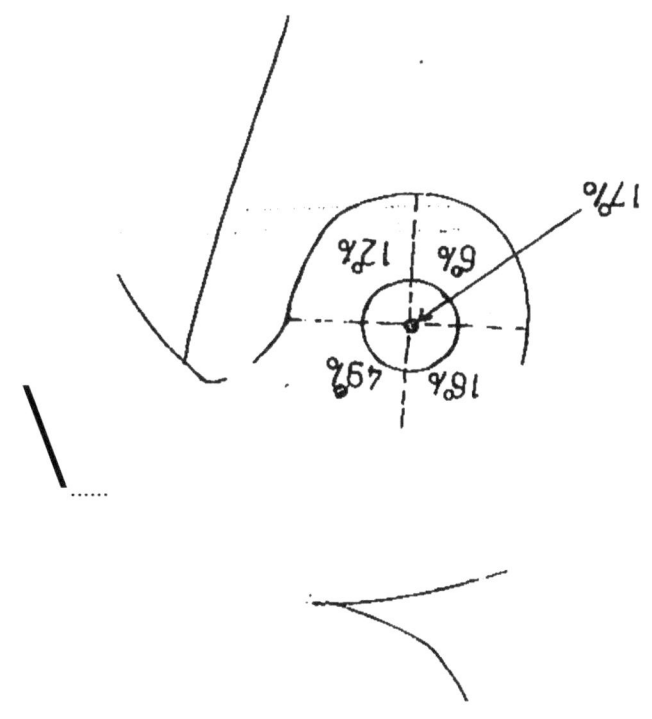

RIA method has been advocated to be the method of choice for the determination of serum prolactin {57} .

1.4.3:Control of Prolactin Secretion :-

The normal concentration of prolactin in human serum is below 400 IU\ L prolactin rapidly cleared from the blood with a half life of less than 30 minutes {58} . Its secretion is controlled by hypothalamic inhibitory and releasing factors .

Thyrotrophin releasing hormone , estrogen and vasoactive intestinal peptide (VIP)can stimulate its secretion (59). Prolactin is released episodically like GH. The highest concentration is found at night and it is dependent on sleep (60) . The concentration of prolactin changes during pregnancy and la ctation and increases progressively up to 10 fold through pregnancy . It is remains elevated during lactation and stimulated by suckling isorders (62) .

1.4.4   :Effects   of   Prolactin :-

The main action of prolactin is the stimulation of lactation . It acts on prepared breast to stimulate growth and support the secretion of milk (63) . The mammary gland is rudimentary in young girls but in the adolescents estrogen GH and adrenal steroids act together to stimulate the growth of the

t system (64) . Alveolar growth is stimulated by estrogen ، GH ، adrenal steroids and prolactin. Following child birth ، prolactin and adrenal steroid are both important for the intiation and propagation of lactation (65).

Prolactin also affects ovaries and testes . It is believed that prolactin control t e jJonaqQtrophin release ، however a high level of PRL cause testiculas_ involution and impotence (66) . In women ، prolactin inhibits estrogen secretion via its mimic of hypothalmus release of gonadotrophin releasing factor ، *this* effect is clear in the postpart(J_tn that is frequently associated wHh menorrhea (67) .

1.4.5: Alterations of Prolactin in Breast Cancer

Hormones may act   at more than one stage in the development of cancer. Abnormal hormonal environment may be responsible for the induction of mammary cancer and may influence the clinical course of the disease (68) . Hormones seem to exert their carcinogenic effect   by increasing the rate of ·proliferation or differentiation of steam or intermediate cells (69) Complete or partial remission of metastatic breast cancer was noted following adrenal_ectomy or ovariectomy or hypophysectomy . The ablatio'! of endocrine glands   removes the source of hormones which are responsible ·for their maintenanc or growth of neoplasm (70) .

Several authors have studied serum prolactin in breast cancer patients but their studies reveal d controver ia.l. _re!?ylt . NrJ.gai et al . and Rose and Pritt.. have reported elevated levels of serum prolactin in postmenopausal women with mammary carcinoma (71 ,72). Bani et al. and Rolandi et. al. reported elevated serum prolactin in breast cancer patients (73,74) .Cole et al. reported elevated serum prolactin in subjects with benign tumours (75).Knishnamoorthy et al. have in estigated serum prolactin and other hormones in patients with breast cancer.They have noted that 7% of premenopausal women have elevated semm prolactin.,while 22% of these patients have elevated prolactin and estradiol . These authors suggested that prolactin involved in the carcinogenesis of the breast (76) . Other authors noted unaltered serum prolactin in women with breast cancer (77,78) .It is obvious from the litreature review that alterations of serum prolactin in breast tumour diseases is not clear:

1.5:Hormo_ne Receptors :-

1.5.1:General Consideration:-

The concept , hormone resptors has limited to such proteins that exhibit d four criteria , they are (79) :

1• Saturability .

2• Affinity .

3 .Specificty .

4. Subsequent biochemical event following hormone receptor interaction.

Receptors may be present either , on the cell membrane (for protein hormone) or in the nucleous (for steriod and thyroid hormones) (80) . Protein hormones act through receptors located on the external surface of the target cell . tc).!nduce changes on the inner surtace of the cell membrane . This change will lead to alterations !n activity within the cell . The afterations transduced by three systems , adenylate cyclase , phospholipase C and tyrosine kinase (81) .

1.5.2:Honnone Receptors in Breast Cancer:-

Hormones act on target cells through the interaction with cellular receptors . In normal cell5 tile receptor concentrat!nn nre regu!at1 d . \ :hereas lll tumour ceils it ls uncontrulied (o2) .A fundamental *trait* uf tun,.:,•Jr r.:t=!ll::. is the decreased dependence on growth fa·:t.ors. Ho'Jlf ,,,.er, snmr tum0•.n: lila hreast carcinoma,are still regulatt:d by a variety of st:erold and rrotr:!!i !.,,-):-rnones (83).This important property is used clinicJ!ly to contro; *coal!* *gro'11th* using antiestrogenic agents .' The presenc.e o · clltmgen rt>ceptr·¹·s *: ppeared* to be essential for tl.!muur re.;por. e tc these £ldcnt Accor·rlif,gly thP.n;e:'l3Urment of estrogen receptors is useful in the designing of hormonal therapy (84). Measurment of estrogen and progesterone receptors (ER , PR) concentrations are useful also as a prognostic factor for the course of the disease and *even* of the pattern of relapse (85). Also the measurment of ER has shown to evaluate the time of survival and it is believed that the presence of estrogen receptors increase the probability of survival (86) . li *is* _obvious from these observations that the roles of ER and PR are essential in the growth of breast tumours (87) .

1.5.3:Prolactin Receptors :-

In contrast to other protein hormones, prolactin intracts *with* receptors in the mammary gland and subsequently without involving any of tile three transducing systems, increases transcription and ultimately translation of massenger RNA which results in the synthesis of *certain* enzymes and milk protein (88) .

Prolactin receptors have been characterized in many tissues including mammary glands, ovary, brain, kidney, liver, testes and prostate (89) . Most of these studies used adioreceptorassay techniques whereas the remaining used immunohistochemical techniques. Prolactin radioreceptor studies revealed that binding of prolactin with receptor are time, temperature , pH and ionic strength dependent (90) . Scatchard plot analysis of the binding revealed frequently a single type of receptors with a high affinity and two types of receptors in somewhat less frequent (91) . Other studies have showed that prolactin receptors are hydrophobic protein (92) .

1.5.4 :Prolactin Receptors in Breast Cancer:-

Shiu et al.were the first authors that had identified prolactin receptors in the mammary glands of pregnant rabbits in 1973 (93) . Following their study , several authors have characterized these receptors (94) • others have reported on the physiological and pathological alterations of prolactin receptors in different conditions (95) .

The involvement of prolactin in facilitating the growth of exprlmentally induced rat mammary tumours has been studied (96) . Several observations have been supported the role of lactogenic hormones in add ion to prolactin in the induction and growth of animal mammary tumours (97). Evidence for prolactin dependence of human breast carcinomas was first reported by salih et al. (98) Barrett et al . observed regression of metastatic breast carcinoma in one patient after hypophesectomy (99). Partridge and Hahne! have found 3 out of 8 tumours of postmenopausal women contained prolactin receptors

(100). Haldway and Friesen, Stagner et al . (101,102) and others (103) have studied prolactin receptors also in breast cancer . In most studies prolactil1 receptors were found in about 50% of the cases .

Bonneterre et al . have investigated the clinical significance of prolactin receptors in breast cancer . They have found that patients with tumours contained prolactin receptors had a better relapse free surviVal although more patients In this group had lumph nodal Involvement and thus recieved adjuvant ghemotherapy (104). It i J>.bYious from these investigations o{ the literature review that the role of prolactin receptors and Its clinical significance in breast cancer patients is not clear .

The Aim of Book :-

The present book deals with the following

1. Alteration of prolactin levels in sera of breast tumours patients
2. Characterization of prolactin receptors in breast tumours
3. Evaluation of the clinical significances of prolactin receptors in the diagnosis and treatment of breast tumours
4. Predication of the relationship between prolactin receptors and serum lipid profile in breast tumour patients

CHAPTER TWO

MATERIALS AND METHODS

Chemicals:-

Alllaboratory chemicals and reagents were of analar grade and were used without further purflcatlon. Dlsodium hydrogen phosphate, citric acid, magnlsum chloride hexahydrate,calcium chloride,copper sulphatepentahydrate, Tris (hydroxymethylamlnomethane), succinic acid, bromocresol green (BCG), sodium azide , Brij 35, and bovine serum albumin (BSA) were purchased from BDH ,UK. Sodium chloride ,sodium hydroxide, glacial acetic acid ,sodium acetate , Sulphosalicylic acid, acetic anhydride, sulfuric acid ,sodium sulfate ,and cholesterol powder were obtained from ORDH,Germany.O-Cresolphthal-ein complexon ,hydrochloric acid, ethanolamine, and calcium carbonate were obtained from Reldal Co. , West Germany . Kits of radioactive prolactin (1251-prolactin) were obtained from Amersham,UK.The specific actMty of 125f-prolactln was 3.2 JJCUri / }JQ protein. Kit for triglycerides determination was obtained from bloMerieux , France.AJl other reagents were of highest purity.

Instruments :-

1 - LKB gamma counter type 1270 Rack .

2 - Beckman Model - 25 Spectrophotometer .

3 - Cooling eentrlfuge type Hettich .

4 - Cooling centrifuge type MLW K24.

5 - Pye - Unlearn pH meter .

6 - Memmert water bath .

7 - Memmert Incubator .

2.1 DeterminatLQ..n of Prolactin L vels in sera of Breast Tumour patients and Healthy Subjects >

2.1.1 Patients:-

Dunng the present study . three groups of breast tumour patients and three gro11ps of healthy subjects were investigate . Group ·r comprised of 25 premenopausal women with malignant breast tumours . Group 2 consisted of 24 postmenopausal omen with malignant breast tumours (11 premenopausal and 15 postmenopausal) .The control groups comprised 21 . and 22 healthy subjects matched with the patients in group 1,2 and 3 respectivly . All patients were newly diagnosed and histologically proven Patients suffered from other diseases were excluded .No patients underwent surgery , chemotherapy or radiotherapy prior to the study.

2.1.2 Blood Sampling :-

Blood samples (7 ml) were obtained from indMduals of all groups by venipuncture and left for 20 minutes at room temperature . After coagulation , sera were separated and kept at -20C until assayed.

2.1.3 Method :-

Serum prolactin levels were measured on samples collected from individuals of all groups by radioimmunoassay. The assay protocol was described in table 2-1 .

2.1.4: Statistical Analysis :-

Students t - test was used to analyze the obtained data.

lable 2-1 :RIA assay protucol ol Serum   Prolactm

	ProlacUn Standards ng/ml						UnknoWn	
	0	5	15	50	100	200	2	
Tubt: m!mbcr	1 2	3.4	5.6	7.8	9.10	11.12	13.14	15 16
Standard Serum	100	100	100	100	100	100		
Unknown serum	—	—	—	—			100	100
rr.1-hPRL	100	100	100	100	100	100	100	100
Anli- Prolactin Serum	100	100	100	100	100	100	100	100

Vortex mix. Incubate 18-24 hours at room temperature(15 - 30 °C).

| Second antrbody reagent | 1000 | 1000 | 1000 | 1000 | 1000 | 1000 | 1000 | 1000 |

Vortex mix, leave at roo111 temperature (15 - 30°C) for at least 5 minutes and centrifuge for 15 minutes at 1500 g.Decant supernatant solutions. .Measure radioactivity in prec1pitatas All volumes in microliter

2.2: Binding of 1251-hPRL to its Receptors in Human Breast Tumours (Malignant and Benign):-

Systematic radioreceptor studies of hormone receptors Include the Investigation of several factors affecting the interaction of the labelled hormone (1251-hPRL)with receptors in the target organ .The choice of the most appropriate amount of protien , optimum pH , temperature, time and salts,are necessary for the studies of these receptors and for the development of radioreceptor assay for prolactin receptors. The details of the methods in this section are based on the incubation of breast homogenate isolated from benign and malignant breast tumour patients with radioactive prolactin(1251-hPRL) at certain temperature, time, pH and subsequently all other parameters should be optimized .

2.2.1: Patients and Tissue Collection :-

All experiments in this section were performed on the three groups of breast tumour patients mentioned In section 2.1.1 .Th y were treated by segmentectomy when the tumour was less than 3 em wide and by total mastectomy when the tumour was bigger or centrally located .Surgery was followed by radiation or chemotherapy •

The excised surgical specimens (1-5 g) were obtained and supplied to our laboratory from Saddam Medical City hospital. The fragment of every breast specimen included in the study were examined histologically and In each case was- confirmed and immediatly rinsed with ice - cold isotonic salln solution.They were collected indMdually in plastic receptcle and stored at -20 C until homogenization .

2.2.2   : Preparation of Breast Tumour Tissue Homogenate:-

Prolactin receptors in the homogenate of breast tumours was prepared according to the method of Bonneterre et al.(104) with some modifications . The frozen tissues of breast were weighted , pulverized finely with a scalpel In petri dish standing on Ice.The pulverized tissues of breast were homogenized in o.o25 M Tris buffer (pH 7.6) with a ratio of 1:3 (w / v) using a manual homogenizer . The homogenate was filtered through five layers of nylon gauze , then centrifuged at 1500 xg for 15 minutes and $4^{o}C$. The supernatant was separated . dMded in aliquots and freezed until the time of the experiments .

2.2.3 :Determination of Protein in Breast Tumour Homogenate:-

Protein content of the homogenate of breast tumours was determined according to the method of lowry et al.(103).The method Is outlined as the following:

1- i ,,:: of Increasing amounts of standard bovine serum albumin (50 , 751 100, 150 . 200 pg / ml) were pipetted In a set of duplicate tubes

2- .1 ml of each sample of homogenate was also plpetted In a duplicate tubes.

3- 5 ml of alkaline solution was added to all assay tubes .

4- The tubes was shaked and allowed to stand at room temperature for 10 min.

5- 0.5 ml of diluted Folin- Ciocalteeu was added to all assay tubes and mixed immediately .

6- The tubes were left at room temperature for 30 min.•

7- The absorbency of the blue solutions were read at 600 nm against the appropliate blank.

8- The standered curve was plotted and the concentration of the homogenate protein was assessed.

Solutions:-

1. Alkaline sodium carbonate solution: (2% Na2C03 In 0.1 N NaOH).

2. Copper sulphate u --potassi m tartrate) solution ;0.5% Cuso,. in 1% Na,K tartrate). This solution was prepared freshly by mixing stock solutions.

.3- Alkaline solution :prepared on day of use by mixing 50 ml of solution 1 and 1 ml of solution 2.

F l!n_-Ciocalteau reagent) : prepared by the dilution of the commercial reagent with an equal volume of water on the day of use.

5- Standard bovtne serum albumin (BSA 0.2 mg /ml).

2.2.4 :Detennination of Tracer Prolactin (1251-hPRL) Concentration :-

' The concentration of the labelled tracer prolactin (1251 -hPRL) was determined according to the method of Morris (106). The method Is outlined in the folloWing steps :

1- 100,ul of 1251-hPRL were pipetted Into assay tubes marked from 1 to 12

2- 100,ul of each standard of unlabelled hormone of concentration ranging from 0 - 200 nglml were pipetted into each tube and according to the assay protocol described in table 2-1.

5- The data in tables 2-2 and 2-3 were plotted as in fig.2-1 .

6- Using the two curves I and II in fig. 2-1 we can get the amount of radioactivity corresponding to the concentration of unlabelled hormone (table 2-4). This was done by drawing a line which intersects with both the curves at the same increament as shown in fig.2-1 .

7- The amount of the standard prolactin was plotted against the amount of the corresponding radioactivity (table 2-4 and fig.2-2).

8- The concentration of the ^{1}251-hPRL was determined from the intercept of the straight line as in fig.2-2 .

2.2.5: Preliminary Test of 1251-hPRL Binding to its Receptors in Breast Tumour Homogenate :-

The binding of 1251-hPRL to breast homogenate preparation was Investigated according to the following :-

1- 50 }Jl of tissue homogenate (400 J.Jg protein) were incubated with 50 }Jl of ^{1}251-hPRL (0.443 nM) and the volumes were completed to 0.5 ml with Tris buffer (pH 7 6), then Incubated at 37 $\overset{o}{C}$ for 3 br.

2- The tubes were centrifuged at 1500 xg for 30 mlnuts and at 4 $\overset{o}{C}$.

3- The tubes were decanted and the precipitates were washed two times With Tris buffer (pH 7.6) .

4- The rims of the tubes were swabbed with cotton pieces .

5- The radioactivity of the precipitate in each tube was counted using a gamma counter. It represents the bound 1251-hPRL .

6- Parallel experiments were used by repeating the steps from 1-5 with the addition of 50 IJI of unlabelled prolactin (4.43 nM) .

Table 2.2: B/F Values Corresponding to Different Concentration of Prolactin Standard Used in Standard Curve .

Concentration of Standard (ng/ml)	B/F
0	0.565
5.5	0.511
15.0	0.390
50.0	0.235
100.0	0.175
207.0	0.146

Table 2.3 : B/F Values Corresponding to Different Amounts of 1251-hPRL Used in the Incubation .

Amounts of 1251-hPRL CPM	Bound Radioctivity	B/F
18999	7050	0.590
28244	10300	0.574
37351	12300	0.491
46095	14180	0.445
55Q11	15520	0.393

Table 2.4: The Mass of the Standard in Nanogram per *ml* Corresponding to a Given Amount of the Tracer .

Bound RadioactMty (CPM x 103)	Amount C nQ/ml)
11	6.0
12	7.5
13	13.2
14	17.3
15	22.0

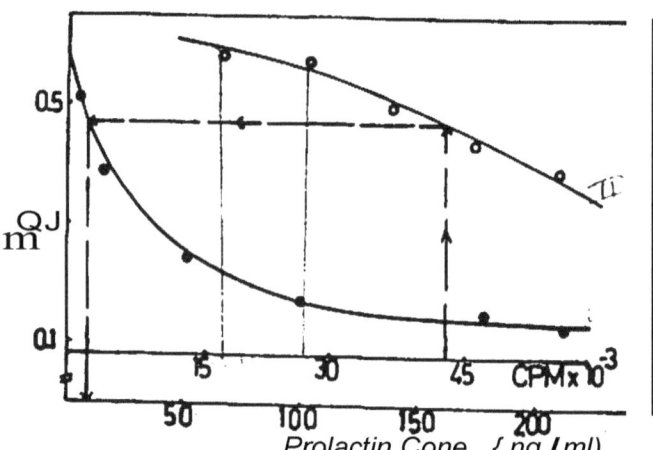

Fig.2.1: Ratio of Bound to Free Radioactivity (BIF) for an OrrJ/nary
Standarr J Cutve, Where Different Amounts of Human Prolactin
StandarrJ were Incubated with Constant Amount .of 12 1-hPRL &
Antibody (I) or, Antibody Incubated with Different Amounts of
12 1-hPRL In the Absence of Unlabelled Hormone (IIj.

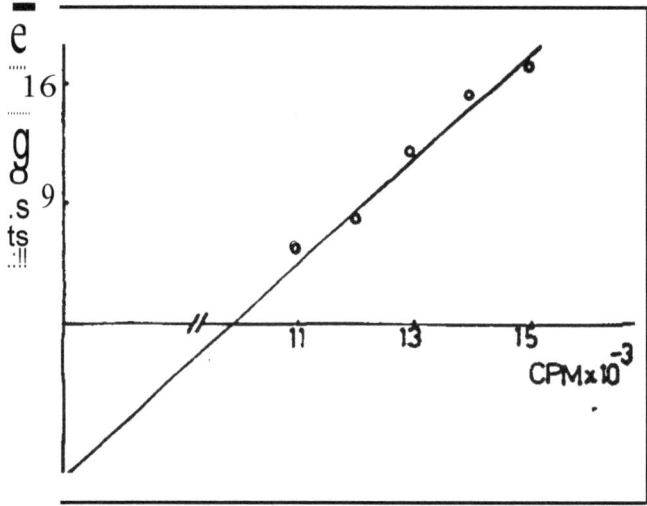

F/g.2.2: A Plot of the Mass of Prolactin StandarrJ Against the CPM of
t2 J-hPRL having the Same BIF from Flg.2.3, Results In a
Straight Line , the Intercept on the Ordinate Corresponds to the
Concentration of 12 1-hPRL In ng/ml

Calculations :-

1-The counted radioactivity in each tube (expressed in CPM) represents the total binding (TB) .

2-The counted radioactivity (expressed in CPM) in the tubes contained labelled honnone and excess of unlabelled honnone (step 6) represents the nonspecific binding (NSB) .

3-The specific binding is the difference between TB and NSB

SB (CPM) = TB (CPM) -NSB(CPM) .

4-SB% can be calculated from the following fonnula:

$$SB\% = (SB/TC > 100$$

Solutions:

1- 1251-hPRL solution : prepared according to the instruction of the kit .

2- Trls buffer (pH 7.6) :prepared by dissoMng the following compounds:

Tris	3.0285 g
$CaCl_2$	2.7748 g',
Bovine serum albumin	0. 1%

in 900 ml of distilled water .The pH was adjusted at 7.6 and the solution was completed to 1 L .

2.2.6: Effect of the Amounts of Prolactin Receptors on the Binding in Breast Tumour Homogenate :-

1- Increasing amounts of protein of breast homogenate (benign and malignant) (50,100,150,200,400,500 pg/50 fll) . were added to a set of six duplicate tubes.

2- 20 f.!l of 1251-hPRL (o.443 nM) was added to each tube and the volumes were   made up to 0.5 ml with Tris   buffer (pH 7.6).

3- The tubes were shaked and incubated at 37°C for 3 hr .

4- The steps 2- 6 of experiment 2-2-5 were repeated.

Calculations:-

The total binding was determined as mentioned In experiment 2-2-5 . It was plotted vs. the amounts of protein in the homogenate.

Solutions:-

As   described In experiment 2-2-5.

2.2.7: The Choice of the most Appropriate Concentration of 1251-hPRL for the Binding with Receptors of Breast Tumour Homogenate:-

1- Increasing   concentrations of   1251-hPRL (0.1-0.8 nM)   were   added to   a set of six duplicate   tubes .

2- 200 pg of protein In the homogenate (50 f-11) was added to each tube and the   volume was made up to 0.5 ml with Trls buffer (pH 7.6) .

3- The tubes were vortlxed and incubated at 37°C for 3 hr .

4- The steps 2 - 6 of experiment 2-2-5 were repeated .

Calculations :-

The total binding was determined as In experiment 2-2-5 and plotted vs. 1251-hPRL concentration.

Solutions:-

Prepared as described in experiment 2-2-5

2.2.8: Effect of pH on the Binding of 1251-hPRL to its Receptors in Breast Tumour Homogenate :-

1- 200 pg of homogenate protein were Incubated in duplicate with 20 pi of 1251-hPRL (0.443 nM) for 3 hr., at $37\,^{o}C$ using buffers of different pH raning from 6-9. The final volume was completed to 0.5 ml.

2- Parallel incubations were performed in the presence of excess unlabelled prolactin (50 fold concentration of labelled hormone) .

3- All incubations were terminated by centrifugation at 1500 xg and $4\,^{o}C$ for 30 minutes.

4- The steps 3-6 of experiment 2-2-5 were repeated.

Calculations:-

1- The specific binding (SB) was estimated as in experiment 2-2-5 .

2- The values of SB% were plotted vs. their corresponding pH.

Solutions :-

1- Trls buffer: prepared as described In experiment 2-2-5.

2- Phosphate buffer : prepared by dissoMng 6.8 g of Na_2HPO_4 & 4.1 g of NaH2P04 in 1 L of distilled water. The different buffer solutions were prepared by mixing accurate volumes of stock solutions and the pH was measured.

2.2.9 :Effect of Temperature on the Binding of 1251-hPRL to its Receptors in Breast Tumour Homogenate:-

1- 50 l of breast homogenate equivalent to 200 pg of protein was added to a set of duplicate tubes containing 20 JJI of 1251-hPRL (0.443 nM) .

2- Parallel experiment was performed to estimate the NSB using 50 folds concentration of unlabelled hormone .

3-   The volume of solutions were made up to 0.5 ml and incubated at $37\,^{o}C$ for

3 hr.

4- The incubation was terminated by centrifugation at 1500 xg and $4\,^{o}C$ for 30 min.

5- The steps 3-6 of experiment 2-2-5 were repeated .

6- The experiment was repeated at _ , 20, 30, 45 ^{o}C.

Calculations:-

1- The specific binding was calculated as In experiment 2-2-5.

2- The SB% wasplotted against the temperature .

Solutions :-

Prepared as described in experiment 2-2-8 .

2.2.1O:Time Course of the Binding of 1251-hPRL to its Receptors in Breast Tumour Homogenate :-

1- 50 pi of breast homogenate (200 /-19 of protein) were pipetted in a set of duplicate tubes contaimng 20 pi of n:. 1-hPRL (0.443 nM) and \he volumes were made up to 0.5 ml with Tris buffer (pH 7.4}.

2- Parallel experiment was performed to estimate the NSB (using 50 folds of unlabelled hormone concentration) .

3- The tubes were incubated at 37°C. At several time intervals (each 45 minutes total and nonspecific binding were withdrawn from incubation medium

4- The steps 3-4 of experiment 2-2-8 were performed and repeated at each 45 minutes interval

Calculations :-

1- The specific binding SB and NSB were calculated as in experiment 2-2-5 .

2- The values of specific binding SB% and NSB% were plotted vs . the time of incubation to yield the time course curves for the association of $_{125}$1-hPRL with its receptors .

Solutions :-

Prepared as described in experiment 2-2-8.

2.2.11 : Effect of Different Halides on the Binding of 1251-hPRL to its Receptors:

1- 50 pi of breast homogenate (200 pg protein) were incubated with 50 pi of 1251-hPRL (0.443 nM)

2- The volume of solutions were completed to 0.5 ml with Tris buffer (pH 7.4) containing 0 01 M of each of the following halides: NaI, NaBr, NaCl and NaF in each assay tube

3- Parallel set of the assay tubes were prepared to detem1ine the amount of NSB

4- The tubes were incubated for 4 hr at 37°C .

5- The steps 3-4 of experiment 2-2-8 were repeated.

Calculations :-

1- The specific binding was determim d as in experiment 2-2-5 .

2- The histogram of the binding was constructed for specific binding in the presence of halides .

Solutions :-

1- Halide solutions prepared In concentration of 0.01 M in Tris buffer (pH 7.4).

    0.3975 g of Nal   in 250 ml of Tris buffer

    0.2575 g of NaBr = = = = = =

    0.1463 g of NaCl = = = = = =

    0.105 g of NaF = = - = = =

2- Other solutions were prepared as described in experiment 2-2-8 .

2.2.12 : Effect of Divalent Cations on the Binding of 1251-hPRL with Receptors of Breast Tumour Homogenate :-

1: 50 pi of breast homogenate (200 pg protein) were added to 20 J.ll of 1251-hPRL (o.443 nM) in duplicate tubes

2- Parallel experiment for each salt concentration was carried out for the estimation of NSB .

3-The volumes of solutions were made up to 0.5 ml with Trls buffer (pH 7.4) In assay tubes contained increasing concentration (10-25 mM) of each of the following salts : MgCl2 , CaCl2 , MnCL2 , CuSO 6H20 .
All the assay incubations were performed at 37^oC for 4 hr.

4-The steps 5 - 6 of experiment 2-2-8 were repeated .

Calculations :-

1- The specific binding was determined as in experiment 2-2-5.

2- A histogram was constructed for the binding of 125f-hPRL to its receptors.

Solutions :-

1-The stock solutions of divalent cations (25 mM) were prepared as the following :-

a- 0.5957 g MgCl2   in 1bf Tris buffer (pH 1:4)

b- 0.6937 g CaCl2 •   =   =   =   =   =

c- 0.7875 g MnCl2   =   =   =   =   =

d-1.5595 g CuS04. 6H20   =   =   =   =   =

2- Other solutions were prepared as in experiment 2-2-8.

2.2.13 : Kinetics of the Association of 1251-hPRL with Receptors of Breast Tumour Homogenate:-

The experiment was carried out to determine the time course of the association of prolactin with its receptor in breast tumour homogenate at different temperatures $_1$ 4, 10, 25, 37 ^{o}C. However the experiment 2-2-10 was 1o\\owed exactly .

Calculations:-

1- The specificbinding (SB) and NSB were calculated as in experiment 2-2-5

2- The following formula was used for the estimation of K_{+1} (107) :

$$\ln [(HR)_\theta / (HR)_\theta - (HR)_t] = K_{+1} [(H)_T (R)_T / (HR)_\theta] \times t$$

where (HR)ll and (HR), are the concentration of hormone- receptor complexes at equilibrium and timet respectivily . (H)r and (R)r are the concentration of the total hormone and the total receptors respectively. K_{+1} is the first order rate constant of the binding reaction .

The slope of the straight line obtained from the plotting of $\ln[(HR)_8 / (HR)_0 - (HR)_1$ vs . t represent: $K_{+1}[(H)r (HR), .t(HR)e]$. Supsequently K_{+1} was determined in four temperatures .

fl.2.14: Thermodynamics of 1251-hPRL Binding to its Receptors in Breast Tumour Homogenate :-

The same steps as mentioned in experiment 2-2-15 were followed exactly , and the experiment was performed at four temperatures i.e, 4, 10, 25, 37 $\overset{o}{C}$.

Calculations:-

1-Thermodynamic parameters of standard state were obtained from Van Hoff plot, the values of natural logarithm of equilibrium constants (affinity constants),Ka obtained at different temperatures.., were plotted vs . the reciprocal values of absolute temperature in Kelvin (1/T) according to the following equation:

$$\ln Ka = \Delta S / R - \Delta H / RT$$

H values was obtained from the slope of th linear relationship of the plot. The change in Gibbs free energy of the standard state (ΔG) was obtained from the following equation: $\Delta G = -RT \ln Ka$, while the standard state entropy change was obtained from the equation :

$$\Delta S = (\Delta H - \Delta G) / T$$

2-Thermodynamic Parameters of the transition state were obtained from Arrhenius plot of ln K vs . 1/T . It gives a linear relationship according to the following equation ·

$$\ln K_1 = \ln A - (Ea / RT)$$

Where A =Arrhenius constant or frequency factor. The value of apparent energy of activation (Ea) of the binding reaction can be determined from the slope of the straight line. The enthalpy of transition state ΔH was obtained from:

$$\Delta H = Ea - RT$$

The transition state free energy change was found by using the following equation :

$$\Delta G = - RT \ln K_1 + RT \ln (KT/ h).$$

Where K is the Boltzmann constant and h is Planks constant The ΔS was found from the following relationship :

$$\Delta s = (\Delta H - \Delta G) / T$$

2.2.15: Determination of thft Concentration ot ProlactiQ Receptors and the Affinity Constant of 125I-hPRL Association with its Receptors in Breast Tumours:-

1- 50 JJl of tumour homogenate (200 jJQ protein) were incubated with increasing amounts of 125I -hPRL (10, 15, 20, 25, 30, 40 Jll. 0.443 nM)

2- Tris buffer (pH 7.4) was added to each assay tube to give a final volume of 0.5 ml.

3- A parallel experiment was carried out to estimate the NSB.

4- All the tubes were incubated for 4 hr, at 37° C in order to obtain an equilibrium state .

5- The steps 3-5 of experiment 2-2-8 were repeated.

Calculations:-

1- The values of bound (B) and free hormone (F) were determined as in experiment 2-2-5 .

2- The values of 1251-hPRL which is bound specifically in nM were calculated using the following formula :

$$Bspac = \frac{Total\ binding\ -\ Nonspecific\ binding}{Total\ counts} \quad Conc.of\ hormone\ in\ each\ assay\ tube$$

3- A scatchard plot was constructed from the values of 8/F versus the values of Bspacific which gives a linear relationship, according to scatchard equation:

$$B/F = Ka\ (\ Btotal.-\ Bspecific.\ )$$

Where Ka: is the affinity constant,

Rtotal : Is the concentration of the receptors (number of binding sites)

Bspecific: is the specifically bound hormone .

2.3: Interactive Effects of Prolactin and its Receptors with Calcium and Serum lipid Profile in Breast Tumour Patients:-

2.3.1: Determination of Calcium Level in Sera and Tissues of Breast Tumour patients :-

Calcium was estimated by a direct spectrophotometric measurment (109) • The method is based on that calcium forms a red complex with 0–cresolphthalein complexone (CPC) at pH 10 – 12 . It is outlined according to the following :-

1- 50 }JI of increasing concentrations of standard calcium were pipetted In a set of test tubes .

2- 50 IJI of sample also pipetted in a duplicate tubes .

3- 2.5 ml of ethanolamine buffer (pH 10.5) was added to each assay tube

4- 2.5 ml of 0-cresolphthalein complexone (CPC) was added also to each assay tube.

5- The tubes were shaked and the absorbancy of the solutions was read Immedlatly at 580 nm .

Calculations :-

The standard curve was plotted and subsequentely, calcium concentration was determined .

Solutions:-

1- 0–Cresolphthalein complexone reagent : prepared by dissoMng 2.5 g of 8 - hydroxyquinoline and 70 mg of CPC in 1 L of 0.19 N HCI (diluted with deionized water) . The solution stored at $4^{o}C$.

2- Ethanolamine buffer: (pH 10.5) prepared by dilution of ethanolamine 3.7 ml to 100 ml with deionized water .

3- stock standard solution of calcium (50 mg/loo ml):1.25 g of dry caco$_3$ was dissolved in 107 ml of 0.77 HCl . The resulting solution was diluted to 1 L with deionized water and stored at 4°C .

2.3.2 :Determination of Serum Albumin :-

Serum albumin was determined using dye binding and according to the method of Doumas et al . (1971) , as modified by Spencer and Price (1977) (110) .The method depends on that bromocresol green forms a blue complex with albumin at acidic pH•The method is outlined as the following :

1- 20 pi of each serum sample (or standard albumin solution) were plpetted in a set of test tubes .

2- 4 ml of working dye solution were added to all test tubes and the solutions were mixed .

3- The tubes were left at 25°C for 10 minutes. The absorbency of the solutions were read at 632 nm against a blank of working dye reagent.

Calculations :-

$$\text{Serum albumin (g / L)} = \frac{\text{Reading of unkown}}{\text{Reading of standard}} \cdot X\,40$$

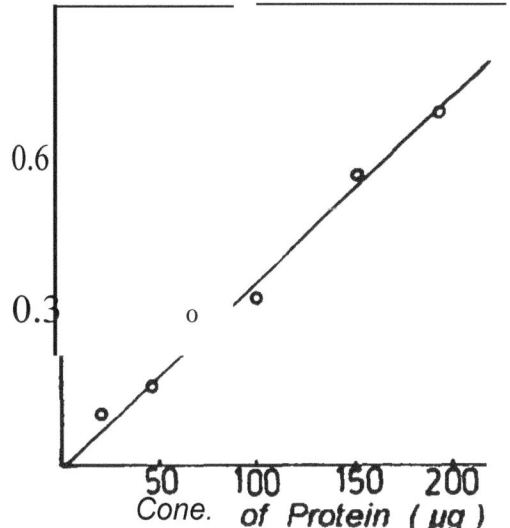

Fig.2.3: Standard curve of *protein determination*

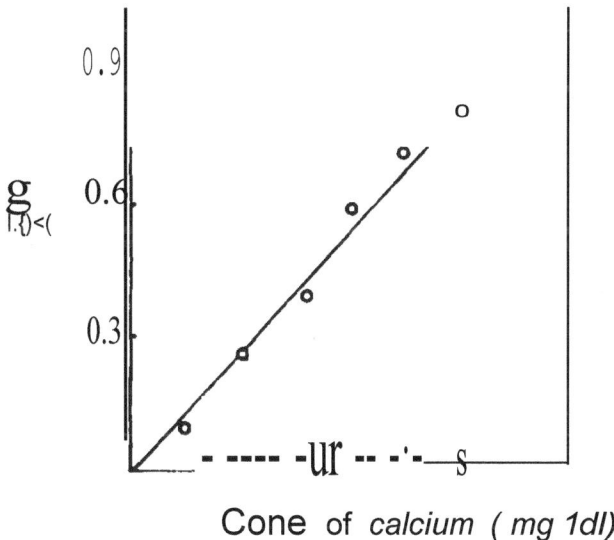

Cone of *calcium (mg 1dl)*

Fig. 2.4 : Standard curve of *calcium determination*

Solutions :-

1- stock succinate buffer (0.5 M , pH 4.1) : 10 g of sodium hydroxide and 56 g of succinic acid were dissolved in 800 ml of water . The pH was adjusted at 4.1 with 1 M sodium hydroxide solution and the volume was completed to 1 L

2- Stock BCG dye solution (10 mM): 1.75 g of BCG was dissolved in 5 ml of 1 M sodium hydroxide solution and the volume was completed to 250 ml with water.

3- Stock sodium azide: 40 g of sodium azide was dissolved in 1 L water.

4- Stock Brij 35 (250 gIL) :25 g of Brij 35 was dissolved in water and the volume was made up to 100 ml with water .

5- Working BCG dye solution (80 pM): To 1 L volumetric flask_, the following solutions were added : 100 ml of stock succinate buffer , 8 ml of stock dye solution , 2,5 ml of stock Brij 35. The volume was made up to 1L with water .

6- Stock standard albumin solution (100 g/L):10 g of human serum albumin with 50 mg of sodium azide were dissolved In water and the volume was completed to 100 ml .

7- Working albumin standards (20,30,40,50 and 60 g/L) prepared by diluting the stock standard with a 500 mg/L solution of sodium azide In water

Note :Since the standard curve is linear In the concentration of the samples studied ,we use 40 g / L standard solution for a single point calibration .

2.3.3 : Detennination of Serum Triglycerides :-

Trigtycerldes were measured by enzymatic method of Fossatl and prenclpe (1982) .The principles of this method depend on the following reactions (111) :

Trigtycerides $\xrightarrow{\text{lipase}}$ **Glycerol** + Fatty acids .

Glycerol+ ATP $\xrightarrow{\text{Glycerokinase}}$ **Gtyceroi-3-P** +ADP

Glycerol-3-P $\xrightarrow{\text{Glycerol-3-phosphate oxidase}}$ Dihydroxyacetone phosphate+ H_2O_2

$2H_2O_2$+p- Chlorophenol+ 4- aminoantipyrln $\xrightarrow{\text{Peroxidase}}$)**quinonimine** + $4H_2O$

The method Is outlined as the following :

1-10 pi of standard glycerol solution (2.29 mM, reagen 1) and each sample were pipetted in test tubes .

2- 1 ml of appropriate working solution was added to all test tubes .

3- The tubes were incubated for 5 minutes at 37 ^{o}C .

4- The absorbancy of the solutions were read at 505 nm agalnts a suitable blank .

Calculations :-

$$\text{Triglycerides in mmol/F} = \frac{\text{Absorbency of sample}}{\text{Absorbency of standard}} \times 2.29$$

Solutions:-

The Kit for 150 determinations contained the following reagents :

1- Reagent 1 (1x8 ml): contained standard glycerol solution (2.29 mmoll/ L) equivalent to 200 mg /1 oo ml of triglycerides (Mw = 875) .

2- Reagent 2 (2x90 ml) : consisted of :

Tris buffer (pH 7.6)	100 mmoi/L
p - Chlorophenol	5.4 mmol/ L
Magnesium	4 mmol/ L

3- Reagent 3 (6X25 ml, lypholized): consisted of:

Amino - Antipyrine	0.4 mmol/ L
ATP	0.8 mmoi/L
Glycerokinase	> 200 u /1
Peroxidase	> 200 u /1
lipase	> 100000 u / L

Working solution: Prepared by reconstitution of 1 bottle of reagent 3 with 25 ml of reagent 2 .

2-3-4: Determination of Serum Cholesterol :-

Serum cholesterol were determined by liebermann - Burchard (1OS) reaction , which based on the ability of cholesterol to be hydrated first , then oxidised by sulphuric acid to link two molecules together and the product can be sulphonated by sulphuric acid or added sulphonic acid . The method is outlined as in the following :

1- 0.1 ml of each sample or standard cholesterol were pipetted in a set of test tubes.

2- 4.5 ml of a mixture of acetic anhydride and sulphosalicylic acid were added to all test tubes .

3- 0.5 ml of concentrated sulfuric acid were added to each tube .

4- The tubes were put in a water bath for 10 minutes at 30 ^{o}C .

5- The absorbancy of the solutions were read against an appropriate blank .

Calculations :-

$$\text{Serum cholesterol in mg}/100\text{ ml} = \frac{\text{Absorbancy of sample}}{\text{Absorbancy of standard}} \times 200$$

Solutions :-

1- Sulphosalicylic acid solution : 5 g of sulphosalicylic acid were dissolved in 100 ml of glacial acetic iacid .

2- The mixture of acetic anhydride and sulphosalicylic acid solution: 35 ml of solution 1 was mixed with 55 ml of acetic anhydride and the mixture was left for overnight . 10 ml of concentrated sulfuric acid was added , then 1 g of sodium sulfate was dissolved and the solution was stored until used .

3- Standard cholesterol solution: 200 mg of cholesterol powder was dissolved in 100 ml of glacial acetic acid _

CHAPTER THREE

RESULTS AND DISCUSSION

3.1: Determination of Serum Prolactin in Healthy Subjects and Breast Tumour Patients :-

Serum prolactin levels were measured in three groups of breast tumour patients matched with three groups of control subjects. Group I consisted of twenty five premenopausal breast cancer patients, group II comprised twenty four postmenopausl breast cancer patients and group III included twenty six benign breast tumour patients.

Table 3.1 and fig.3.1 show the results of this study. In women with benign breast tumours, the level of serum prolactin is not altered significantly as compared with that of normal subjects. A significant increase In serum prolactin is observed in premenopausal cancer wom n (P<0.01), however, nine of the twenty five patients (36%) of this group exhibited high level of serum prolactin. A significant increase in serum prolactin is also observed In the group of postmenopausal breast cancer women but with low extent (P< 0.05) and six of the twenty four patients (25%) of this group showed high level of serum prolactin.

The increase of serum prolactin level in the two groups of breast cancer patients and the unalteration of that level in the group of benign tumour patients confinn the results of the previous studies, which revealed. several abnormalities in serum prolactin of malignant breast tumour patients (76). The apparent increase in prolactin levels in the two groups of the malignant tumour patients may be due to the increased sensitization of prolactin- secreting cells of the pituitary by estradiol as the latter reported to be increased in breast cancer patients (103), but this hypothesis needs further Investigation to be well understood.

Table .3-1 :Serum Prolactin (ng / ml) in Normal Women, Patients with Benign Breast Tumours Premenopausal and Postmenopausal Women with Breast Cancer.

	No. of Cases	Age (year)	Serum Prolactin
Group 1	25	$33 + 4.1$	$30.5 + 10.2$ [a]
Control	21	$31 + 4.3$	$20.5+ 9.1$
Group 2	24	$54+ 5.2$	[b]
Control	21	$53+ 4.7$	$22.5+ 8.7$
Group 3	26	$45 + 7.5$	$25.Z + 11.8$
Conrol	22		$22.3+ 7.9$

a: $P < 0.01$

b: $p < 0.05$

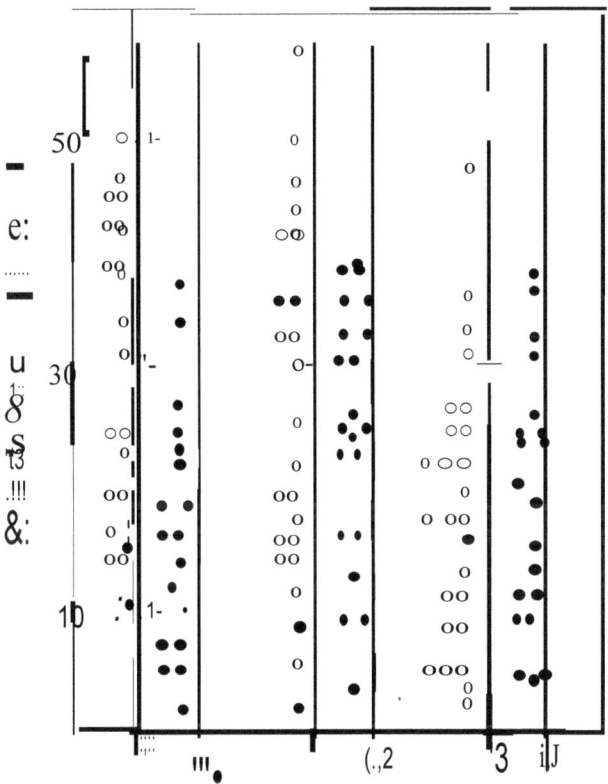

Fig 3.1: 0/strlbut/on of Serum Prolactin (nglml) In Normal Women ,
    Patients   with Benign Breast Tumours , Pre and Postmenopausal
    Women With Breast Cancer .

P1 , C1 : Patients and Control In Group 1 Respectively .

P2 , C2 : Patients and Control in Group 2 Respectively

P3 , C3 : Patients and Control In Group 3 Respectively .

The high levels of serum prolactin in the two groups of the breast cancer patients may Indicate that prolactin was involved in the carcinogenesis of breast tumours at least in some women with breast carcinoma (112). Since prolactin plays a major role In the function of the breast , the implication of the hypersecretion of this hormone may be seen in the increased sensItMty of breast tissue to carcinogenesis. It was also observed that an Increase in serum prolactin was found to specifically stimulate some of enzymes involved in lipid and carbohydrate metabolism in cancerous breast (76). So the elevated levels of serum prolactin in some of the breast cancer women may be useful In the diagnosis and management of this disease particularly In premenopausal women.

Binding of 1251-hPRL with its Receptors in Human Breast Tumours (Malignant and Be_nign):-

3.2: Preliminary Test of 1251-hPRL Binding to its Receptors in Homogenate of Breast Tumour patients:-

As mentioned In section 2-2-5 three groups of bre9st tumour petients were Investigated for the presence of prolactin receptors after removal of the tumours .400 JJQ of tissue protein and 20 IJI of tracer prolactin (0.443 nM) were Included in each test In the presence or b P.nr. of 20 fold concentration of unlabelled prolactin . A tumour was considered to contain prolactin receptors if the specific binding $\leq$ greater than 0.8% , the latter value was used as a criterion for the presence of prolactin receptors because when the assay was performed on denatured homogenate protien , the difference between the binding in the absence or presence of unlabelled prolactin was always less than 0.8% (104).

In group 1, prolactin receptors were found In eleven out of twenty five tumours (44%) of the premenopausal breast cancer patients .The range of the specific binding was found to be 0.8-18% . In group 2 , prolactin receptors were detected In eight out of twenty four tumours of postmenopausal breast cancer patients (33%) Investigated .The range of the specific binding was found to be 0.8- 16.5%. However in group 3, prolactin receptors were existed In two out of twenty six (8%) benign tumours studied and the extent of the specific binding was 0.8- 7.5% .

The results obtained is cosistent with those reported previously (104). However several authors also obtained broad ranges of variation In prolactin receptors in the cases studied particularly those of malignant cases (about 10 - 49% of the cases). The differences may be attributable essentially to the variation In the labelled hormone concentration, the conditions of homogenat preparation ,the amount of protein used and other Incubation conditions (103) .

3.3: Effect of Protein Concentration on 1251-hPRL Binding with its Receptors in Breast Tumour Homogenate:-

To determine whether the specific binding was proportional to the amount of protein of the receptors actually present In the incubation mixture, Increasing amounts of homogenate protein were incubated with either tracer prolactin alone or with unlabelled hormone added . Fig.3.2-A shows an Increase In

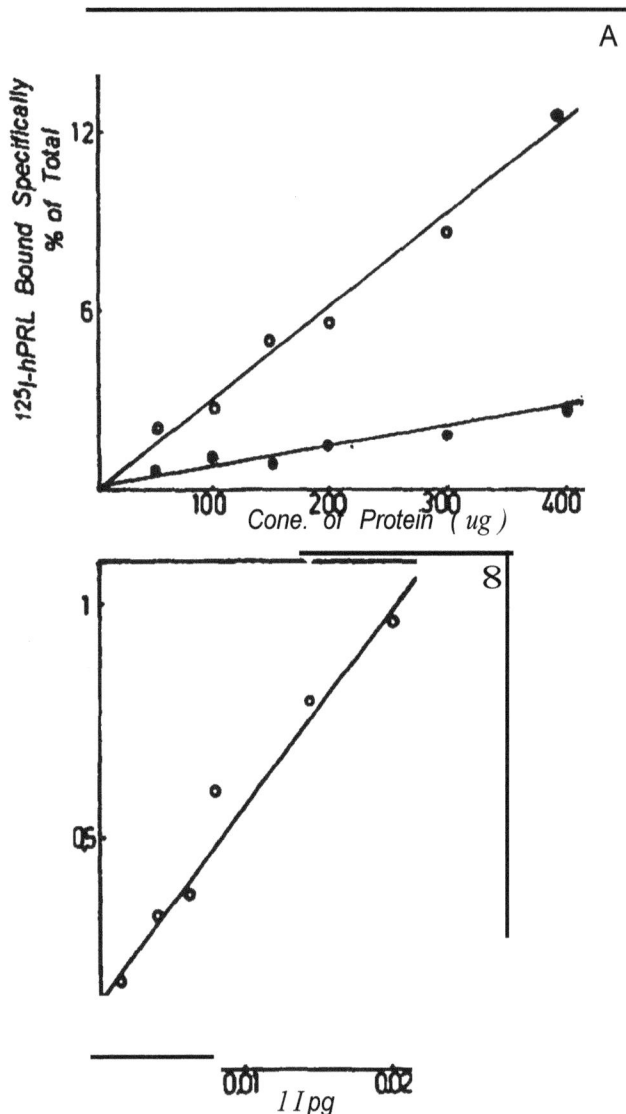

Fig 3.2: A: Incubation of Increasing Amounts of Malignant Breast

Tumour Homogenante with Labelled Prolactin.

B: An Inverse Plot of Data In A

o: Specific Binding

• : Nonspecific Binding

the percentage of 1251-hRPL bound specffically to protein of breast tumour homogenate as the latter increases in the incubation mixture Thus the specific binding Is increased linearly whereas , nonspecific binding is also increased linearly but with low extent .These results indicate that 1251-hPRL binding are principally depended on the amounts of homogenate protein in the reaction mixture (113) . Normally 200 JJg of protein was used per incubation in the subsequent experiments .

A IIneweaver - Burl< plot (114) , can be constructed from the data shown In fig 3-2-A ,and as illustrated in fig 3-2-8,1251-hPRL binding With breast tumour protein gives a straight line.The Y - intercept of the plot represents the Inverse of the maximum amount of tracer prolactin that could be specifically bound to an infinite amount of receptors.This type of plot was suggested by Haro and Talamantes (107).

3.4 : Effect of 1251-hPRL Concentration on the Binding with its Receptors in Breast Tumour Homogenate :-

saturabllIty Is one of the four criteria which can fulfil the concept of true receptors (79) .The experiment was carried out using 200 pg protein and increasing concentration of 1251-hPRL in order *to* evaluate this characterlstic.Fig.3.3 shows that specific binding of ^{1}251 -hPRL with breast tumour protein is a saturable processs, but complete saturation ,however is theoretically never reached ,unless the amount of prolactin used reaches infinity (93) .

From fig.3.3 , it is obvfous that breast tumour protein was saturated with labelled prolactin when the amount of the latter In the Incubartor mixture was equivalent to 20 J.ll (0.443 nM) therefore ,the latter amount o1 tracer prolactin was used in the subsequent experiments.

3.5:Effect of pH on 1251-hPRL Binding with its Receptors in Breast Tumour Homogenate:-

The analysis of the effect of pH on labelled prolactin binding with breast tumour protein is shown In g.3-4 .The optimum pH was found to be 7.4 for the two types of tumours ,malignant and benign. The profile observed *in* rhts fig. shows also that the maximal binding occurs over a relatively narrow pH range.However there Is a sharp decline In the specific binding beneath the optimal pH and on the other hand the labelled hormone precipitates ti"om solution at pH values less than 6.5 under the incubation conditions used.

These results are in agreement with those reported previously .Shiue and Friesen (93) obtained the pH 7.3 as the optimal for the binding of prolactin with mammary gland receptors.Al-Khayat (89) reported that the optimal pH of p lactin binding With prostatic homogenate was 6.8, while Al- Mahdawt(#&)-showed that the optimal pH of the binding was 7 6.

According. to the results of this experiment • the buffers in al experiments were adjusted at 7.4 and used as optimal pH

3.6:Effect of Temperature on 1251-hPRL Binding with its Receptors in Breast Tumour Homogenate:-

Fig. 3.5 shows the effect of temperature on labelled prolactir' htn'ltr5] with breast tumour protein of both malignant and benign sper'''''·   -

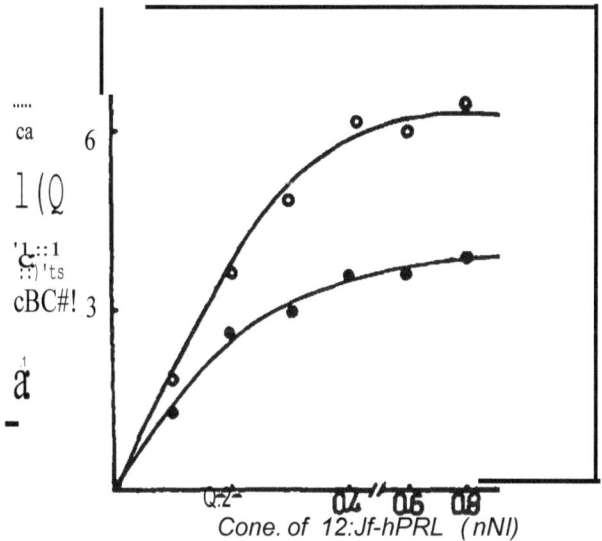

Fig 3.3: Effect of Different Concentration of 1251-hPRL on the Binding with Breast Tumours Homogenate . All Details are Explained In the Text . o : Mal/gnat Tumour Homogenate ، • : Benign Tumour Homogenate .

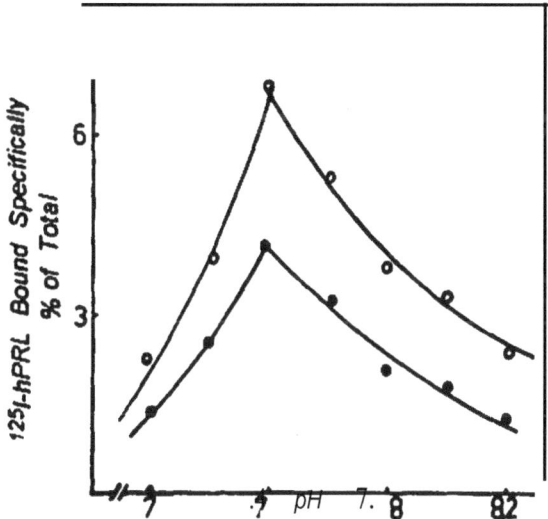

Fig 3.4: P'H Dependency of 12'1-hPRL Binding with Breast Tumour Homogenate. All Details are Explained In the Text. D: Mallgnam Tumour Homogenate, • : Benign Tumour Homogenate .

tJ1nd111g profile demonstrated was rdent1cal tor both types of tumours an .- veals temp rati..il e dept: r.dency rhe specific binding was mereased as till tt: mpemture increased. rt:achmg a plateau ;.)t 37 ^{o}C , then declinth1 sharply, whereas the nonspecific binding was not temperature dependent. Since specific bmding reached , max1mal equilibrium at 37 ^{o}C under the conditions of incubation investigated, accordingly thiS temperature was used as tha optimum temperature in all subsequent incubations

3.7: Time Course of 1251-hPRL Binding with its Receptors in Breast Tumour Homogenate:-

To examine the characteristics of the binding, experiment was performed to study the stability of the in vitro system . Flg.3-6 shows the time course of the binding reaction a1ter Incubation of tracer prolactin with or without unlabelled hormone. The specific binding for the two types of tumours increases linearly for the first 2 hr . reaches a plateau level at 4 hr. and remains constant for the next 4 hr . However the percentage of the specifically bound tracer prolactin was found to be 15 at the equilibrium plateau • while for nonspecific binding being approximately 5 at all time intervals examined . Since equilibrium was reached at 4 hr . suitably this interval was used in all subsequent studies .

This results is not in agreement with those obtained previously. However Shiue and Friesen reported that after 3 hr.• the specific binding of prolactin with receptors of mammary gland reached equilibrium (93).

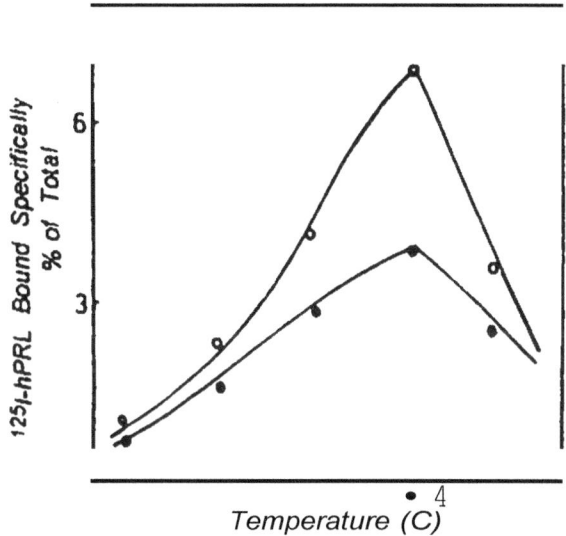

F1g 3 5 ·Effect of Temperature on the Binding of 12'1-hPRL with
Breast Tumour Homogenate   o : Malignant Tumours ,
• ·Benign Tumours

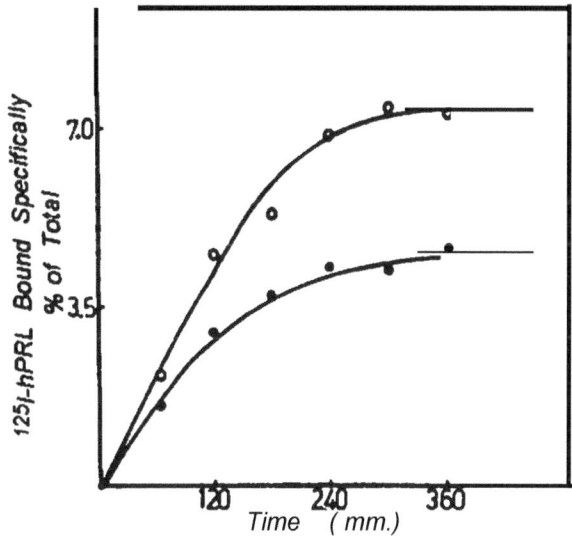

F1g 3.6: Time course of 1251-hPRL Binding with Breast Tumour
Homogenate. o : Malignant Tumours ,• . Benign Tumours .

All Details are Explained in the Text

This finding may be attributed to the fact that malignant and benign tumours of human were Included in our study , while other authors used normal tissues Isolated from rats .

3.8: Specificity of 1251-hPRL Binding with Breast Tumour Receptors:-

1251-hPRL specificity of the binding sites was studied in compititlon experiment with unlabelled prolactin.As shown In fig.3-7, when Increasing amounts of nonradioactive prolactin were Incubated , a progressive decrease in the percentage of 1251-hPRL bound to breast tumour receptors was found to be 12 and 8 for malignant and benign tumours respectively, progressively decreased to 4.5 , when the concentration of the unlabelled prolactin equivalent to 20 fold of that_, of tracer prolactln.The latter value remained constant whatever 'the unlabelled prolactin Increased In the Incubation medium. The 4.5% value was considered to represent nonspecific binding ,and was substracted from the percentage of 1251-hPRL binding ·In the subsequent studies.

The displacement of 1251-hPRL by unlabelled hormone confirms the specificity of the breast tumour binding with receptors , which Is one of the fundamental criteria of the true receptors (79) .

3.9:Effect of Halides on 1251-hPRL Binding with its Receptors in Breast Tumour Homogenate:-

Different halides of sodium were investigated for their action on the binding of prolactin with breast tumour homogenate.The results are Illustrated

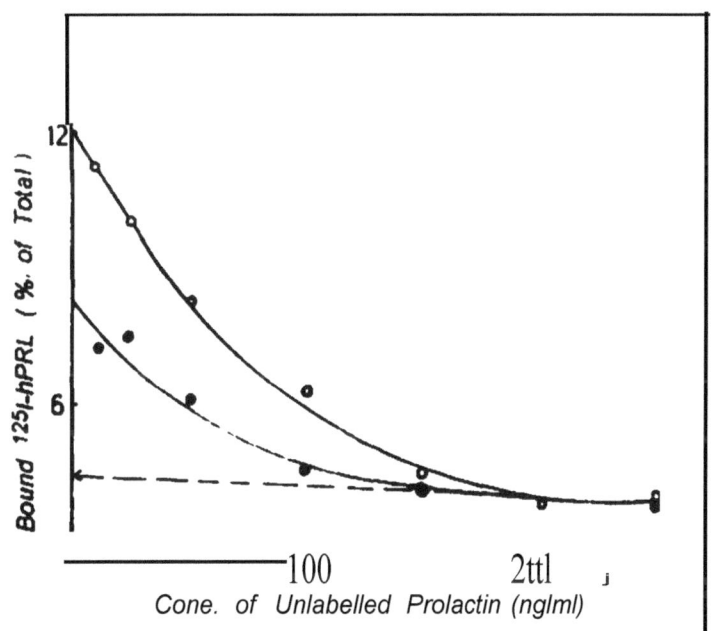

Fig 3.7 : *Competition for 12St-hPRL Binding In Breast Tumour Homogenate.   o: Malignant   Tumours,· Benign   Tumours.*

All   Details   ere   Expllained   in   the   Text

in fig.3-8.The presence of sodium halides in the Incubation medium tends to promote the binding of prolactin according to the following sequence :

$$NaF_{:::} > NaCl > NaBr > NaI$$

This frequency of binding stimulation may be demonstrated thereby the increase of hydrophobic interactions which Is essentially involved in the binding process (116) .

3.1O:Effect of Divalent Cation on the Binding of 1251-hPRL with its Receptors in Breast Tumour Homogenate :-

The Importance of Ionic environment for the binding of prolactin to breast tumour protein is shown in fig.3-9. Divalent cations appeared to enhance the binding reaction at low concentration while, at high concentration of all salts, inhibition of the binding has occured . Among all the cations studied calcium was the most importan for the stimulation of prolactin binding. However at concentration ranging from 10 mM to 20 mM $\cdot$ $CaCl_2$ increases the binding three folds,so In all our studies , 20 mM of calcium was included In the incubation medlum.The frequency of the stimulation by divalent cations is according to the following :

$$Ca \cdot 2 > Mg \cdot 7 > Mn+7 > Cu+2$$

The result obtained shows that calcium Is essentially involved In the binding process of prolactin with breast tumour proteins .This evidence may suggest that prolactin binding Is a calmodulin dependent process (118). Our results in accordance with that obtained by several authors (93),and Is different from those reported by others(89).The variation between these results

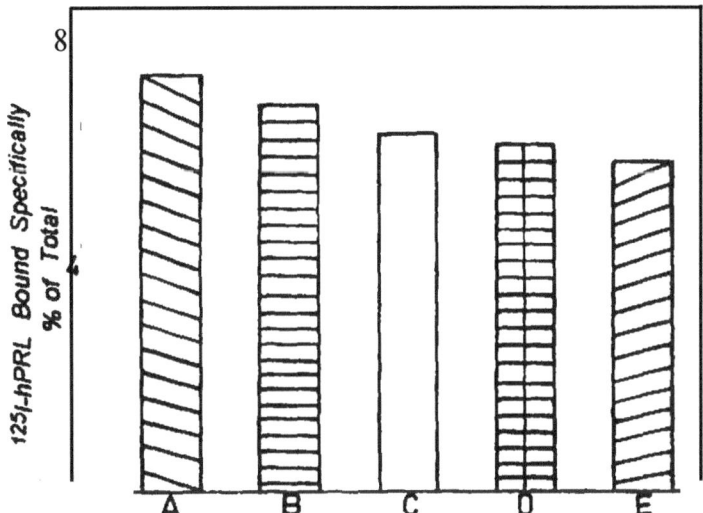

*Fig 3.8: The Effect of Different Halides on 1251-hPRL Binding with
Ma/lgnanr Breast Tumour Homogenate • A : NsF, B : NaCf
C : NaBr. D : Nal. E : Without Addition of Halide •*

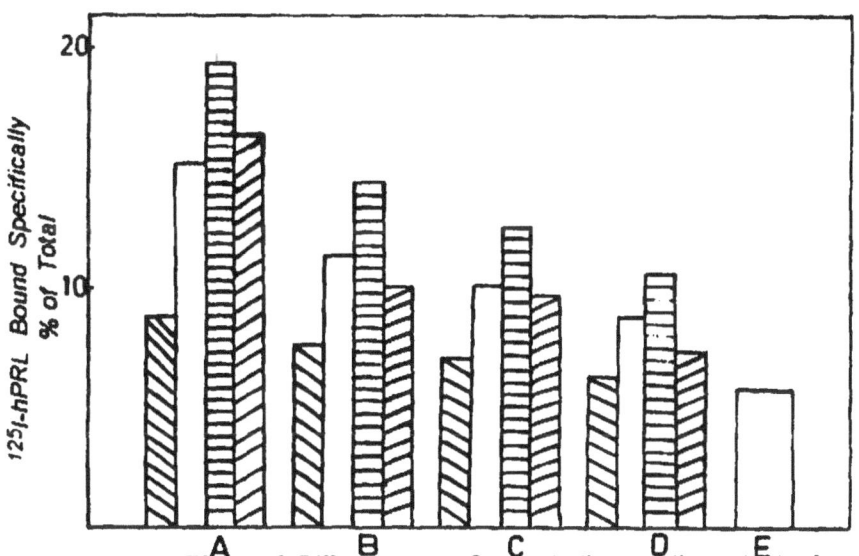

*Fig 3.9 : The Effect of Different sans Concentration on the extent of
12SJ-hPRL Binding with Malignant Tumour Homogenate .A : C&C/2,
B: MgC/2, C: MnC/2, D: CuC/2·ISSJ;10 mM, 0:15 mM, E3
;20 mM. £221 ;25 mM, E : Without Addition of sans .*

and that obtained by others may be ascribed to the difference of the tissue studied (103) .The Involvement of calcium in the binding of prolactin, with breast tumour protein reminded us to estimate the concentration of this cation in tissues and sera of breast tumour patients .

3.11:Kinetics of 1251-hPRL Binding with Breast Tumour Receptors :-

Since the binding of prolactin with breast tumour receptors Is time and temperature dependent, the time course of binding was further investigated at four temperatures (4 ,15 , 25 , 37 $\overset{o}{C}$) .The association rate constant (:V of prolactin with receptors of malignant breast tumours was estimated at the four temperatures studied .

The time course data shown in fig 3-10 fits pseudo first order Kinetics for the association of 1251-hPRL with its receptors in breast cancerous tumours.The values of K+1 at the four temperatures are tabulated In table 3-2 .Results show that the highest rate of association occurs at 37 $\overset{o}{C}$ whereas the lowest rate occurs at 4 $\overset{o}{C}$. When the reaction temperature was increased from 4 $\overset{o}{C}$ to 37 $\overset{o}{C}$,the value of the association rate constant increased approximately 5 folds .This result is similar to those obtained for the binding of prolactin with receptors of liver (107) and prostate (89). In general the slow rate of prolactin binding with receptors of malignant breast tumours suggests that the binding is a diffusion controllled process (93) .

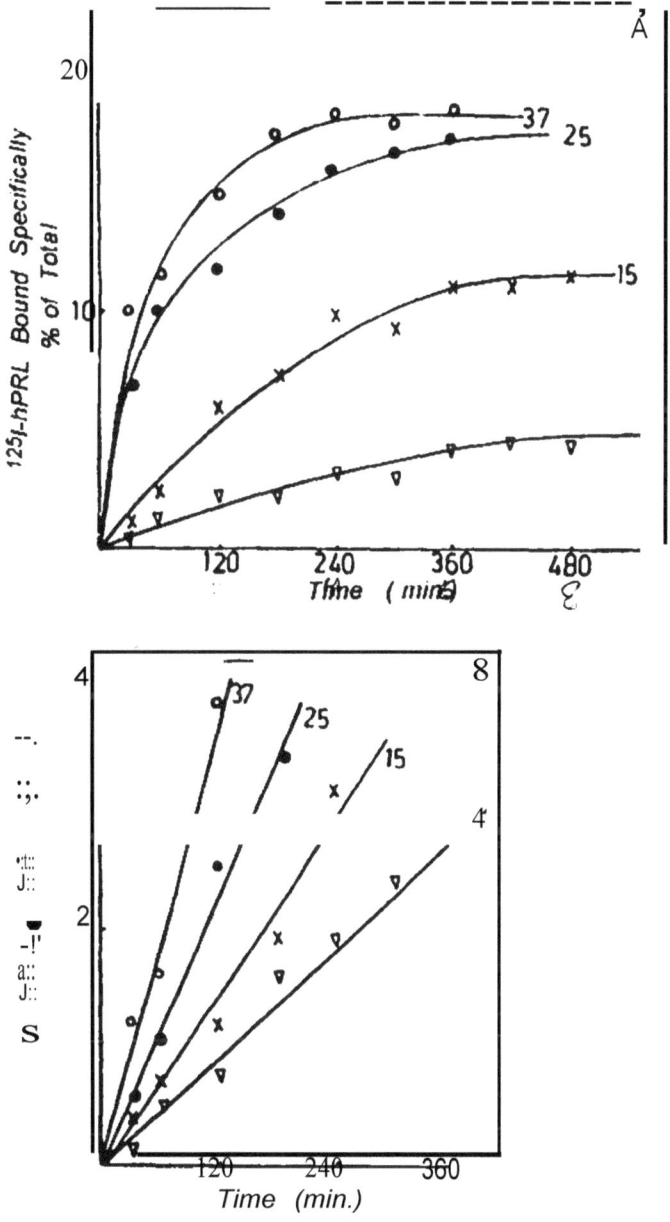

F1g 3.10 A ·Time Course ·for the Association of 125J-hPRL w1tll
Receptors of Malignant Breast Tumours at Different
Temperatures B ·Pseudo first Order Pint of the
Binding Data in A All Details are 6.piained in the
Te:..t

3.12:l_t}!!rmo. ynamics of 125I-hPRL Binding with Receptors of _Malignant Breast Tumours :-

In order to fuffif thF.! chnractm·: ation of prolactin receptors In malignant breast tumours,the thermodynamic parameters of the standard state were estimated usfng the P.qui!ihriun'l ffinity constants . These constants were measured at four t mperatur s (_ , 15 • 25 , 37 $\overset{0}{C}$), and seemed to be temperature dependent (Table 3- 3).Vant Hoff plot revealed a constant positive enthalpy of the standarr! state at the temperature range studied, suggP,st!ng endothermic reaction (Fig.3-11)The high values of entropy of the standared stat suggP.s s that prolactin binding with its receptors was entropir.R 1 1 driven and th9 ..e:tc!!n"" was essentially associated with conform3tional changf. ('J 9).

The re lJits obt !n d is of interest , since prolactin binding In our system revealed a constant enthalpy , in contrast to most protein and polypeptide hormones which were showed temperature dependent enthalpy for their binding with receptors (·118).

The thermoctyr,amic parameters of the transition state formation of prolactin - receptor complex were determined using A!!flenlus equation (Fig 3-12) • The posf•lve Glbbes frAe energy indicates that the formation of the transft.lon state requ'"ed In put of anergy .The lower values of entropy indicates that t.hFJ drMng fon:e 'Nas h£ enthalpy hence , the reaction was not accompt?. ied by r.r.-r.formatinnal ch nr;"!.

lable   3 - 2 · The Association   Rate   Constants $(K+_1)$ for the Bindin{;   of 1251-hPRL   to   its   Receptors   in   Malignant   Breast   TumoUis   at   four   Temperatures

Temperature $(\overset{o}{C})$	$K_{+1}(\ M^{-1} x\ min^{-1} x\ 10^9\ )$
4	1.21+o.12
15	3.50 + 0 22
25	6.50 + 0.21 .
37	9.35 ;-0.28

Table 3-3 ·Thermodynamic Parameters of Standard and Transition States of 125J-hPRL   Binding with Malignant Breast Tumour Receptors

Tem. $(\overset{o}{c})$	Ka $\{\ M-1)(10^{10})$	/jG (KJ/mole)	H	11 (J/mole.K)	Ea (KJ/mole)	H*	S• (J/mole .K)
4	0.41	-51.0	5824	394.9	46.48	44.17	89 16
15	1.21	-55.5	=	394.4	=	44 10	9105
25	2.20	-58.8	=	394.1	=	43. 11	90.93
37	3.26	-62.3	=	388.8	=	43.90	87.38

68

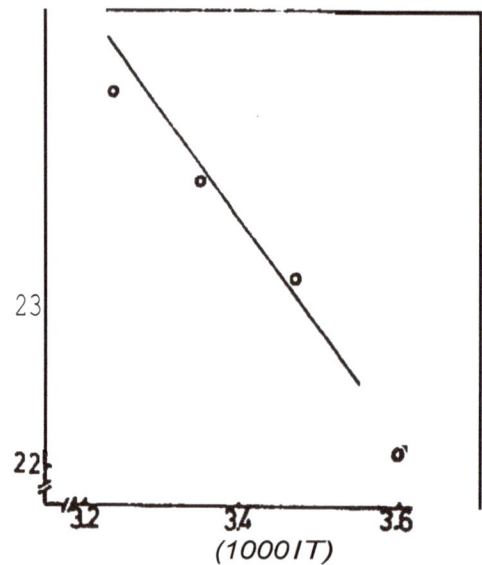

Fig 3.1 J - **"Plot** *of 1251-hPRL Binding with Receptors of Malignant Breast Tumours   All Details are Explained in the Te;(t*

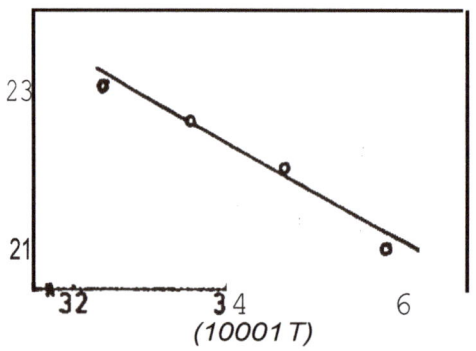

F1g 312 ·ArriJenius Plot of 1251-/JPRL Binding with Receptors of Malignant Breast Tumours   .4/i Detei'':5 a:e &piaintJd ir. ll e Toxt

3.13: Determination of the Concentration of Prolactin Receptors and the Affinity Constants of 1251- hPRL Binding with Breast Tumour Receptors:-

Prolactin receptor concentrations were measured in breast tumours exhibited specific binding with prolactin {more than 0.8%) as described In section 3-2. The affinity constant of the binding of prolactin with its receptors were also measured in the same tumours (Fig.3-13). The results are summarised in table 3-4 . Prolactin receptors were increased significantly (P<0.001) in group 1 {premenopausal) patients as compared with group 3 which was consisted of patients With benign tumours. In group 2 (postmenopausal), Prolactin receptors were also increased significantly (P<0.01) in respect to group 3 but with low extent . The affinity constants of the binding were not altered significantly in the three groups of breast tumour patients suggesting a similar affinity of prolactin for its receptors in these groups .There is apparently no correlation, in breast tumour patients , between their circulating prolactin levels and the presence of its receptors .

Prolactin is believed to be involved in mammary tumourtgenesis (96) and this hypothesis was documented at least In part of our project (as described in section 3 - 1) . Prolactin has been shown to Induce several unfavorable effects in the human. Thus , Malarkey et al . {120) have demonstrated in vitro the positive effect of prolactin (as well as of growth hormone)on the growth of cell from breast carcinoma of a hyperprolactinemic patients .Holt- kamp et al .(121) have found a 30% incidence of

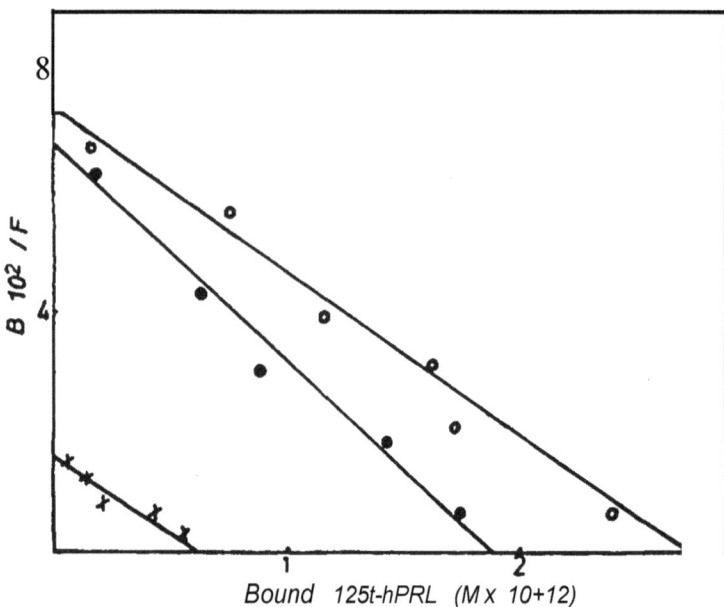

Fig 3.13: Scatchard Plot of 1251-hPRL Blndi.'1rJ wnh ns Receptors In
the Breast Tumours of the three Groups Studied .
o: Malignant Tumours of Premet Op811S81 Patient.<:
• ·Malignant rumours o Postmenet(IBIIsPI Patients
x : Benign Tumours

Table. 3 - 4 :Radioreceptor Assay for Determining Prolactin Binding and Affinity Constants in the Three Groups of Breast Tumour Patients Studied .

	No. of Cases	Age (year}	Binding Capacity (fmole / mg)	Ka $(M^{-1})(10^{10})$
Group 1	25	33 + 4.5	a	2.74 + 0.31
Group 2	24	54+ 5.2	b	3.41 + 0.52
Group 3	26	45 + 7.5	1.55 + 0.97	2.66 + 0.29

a:$P < 0.001$

b:$p < 0.01$

hyperprolactinemla In patients with metastatic breast cancer , with highest incidence in patients with familial type of breast cancer , especially with extremely aggressive disease and short disease free interval .

The cellular responses for prolactin are dependent upon its Interaction with specific receptors (88) . The number of prolactin receptors In a particular tissue can also vary with the hormonal environment of the tissue (95) . The ability of prolactin to stimulate tumour growth may therefore depends less on serum levels , and more on the number of receptor sites in the tumour tissue (103) •

Our study revealed that prolactin receptors were detected in some tumours and there was no correlation between these receptors and the levels of the hormone in serum. This is in contrast to induce mammary tumours in rats, In which , prolactin levels were increased, the tumours appear to be dependent on the hormonal level , and prolactin receptors were present and subjected to hormonal as well as pharmacological modulation (122) . The appearance of prolactin receptors In about 40% of the tumours studied of the two groups of breast cancer patients (malignant tumours) , and the low abundance In group III (8% of benign tumours), moreover the significant increase of these receptors In the malignant groups Indicate that prolactin receptors may be used as a diagnostic factor in breast cancer . In addition to that , these receptors may be used also for the assessment of relapse free period or (and) the over all survival , as the case of estrogen and progesterone receptors (86).

Previously , prolactin receptors have been Investigated In breast' cancer and revealed controversial results . Holdaway et al .(101) suggested that 20% of the 41 cases of breast tumours studied contained prolactin receptors . Partridge et al .(100) showed that 33% of the 9 cases Investigated contained prolactin receptors . Payrat et al . (123) demonstrated the presence of prolactin receptors in 49% of the 72 cases studied . Bonneterre et al .(124) deduced a significant binding of prolactin In 46% of the 92 cases Investigated .Turcot- Lemay et al. (125) showed that prolactin receptors existed In 36% of the 343 cases studied .However the apparent variations In these studies may be attributed essentially to several factors related to the assay conditions, particularly the source of prolactin (103) .This interrupted factor was excluded from our study , since we have used a highly purified human prolactin •

Prolactin receptors in breast tumours have been shown to be modified and affected by several factors . However an early trail of bromocrtptlne effect on breast cancer tumours led to the conclusion that Inhibition of prolactin would have no benefit in the treatment of advanced breast cancer (127). Since this agent Inhibits the secretion of prolactin but can Increase the level of growth hormone consequently , the latter Is capable to bind prolactin receptors leading to the same tumourlgenlc effect as the case of high level of prolactin (126) .Steroid hormones and their receptors have been shown to affect and modify prolactin receptors . Experimentally estradiol stimulates prolactin receptors biosynthesis In vitro and In vivo (128) Conversely prolactin stimulates the appearance of estrogen receptors in human breast cancer cells (129),on the other hand anti-prolactin treatment leads to a decrease in estrogen receptor levels In rat mammary tumours (130) .Bonneterre et al .confirmed a significant correlation between estrogen receptors and prolactin receptors (104) .However these observations indicate that the measurment of prolactin receptors may become more effective in the management of breast cancer if steroid hormone receptors are measured consectively .

3.14 : Determination of Calcium in Sera and Tumour Extracts of Breast Cancer Patients :-

We have shown that in vitro study of ^{125}I-hPRL binding with breast tumour receptors is principally depended on the concentration of calcium in the incubation medium . At a concentration of 0.02 M of calcium . the binding was inceased 3 fold This observation evoked us to investigate the concentration of calcium in sera and tumour tissue extracts of the three groups of breast tumour patients . Patients which have metastasis in any site were excluded from our study .The approach of such study Is to investigate the factors mediated the tumourigenic effect of prolactin and Its receptors in breast tumours .

Results obtained in table 3 - 5 show that serum calcium was Increased significantly ($P < 0.01$) in the two groups of cancerous patients as compared with that of healthy subjects . whereas in the group of benign tumour patients, serum calcium remained unaltered . Also the concentration of calcium in the tumour tissue extracts was not flucatuted significantly in all breast

Table . 3 - 5 : Serum Calcium and Albumin and   Tissue   Calcium   in
Breast Tumour Patients of   the Three Groups Studied ,
Compared   to the Healthy   Subjects .

	No. of Cases	Age (year)	S.Calcium (mg/100 ml)	Tissue Calcium (mglg tissue)	S.AJbumin (mg/100 ml)
Group 1	25	33 + 4.1	a 10.12 + 0.92	798 + 360	4.12 + 0.70
Control	21	31 + 4.3	9.2 + 0.61		4.4 + 0.73
Group 2	24	54+ 5.2	a 9.93 + 0.85	920 + 392	4.11 + 0.77
Control	21	53+4.7	9.14 + 0.47		4.23 + 0.63
Group 3	26	45 + 7.5	9.37 + 0.45	987 + 411	4.06 +0.83
Control	22	47 7.3	9.28 + 0.53		4.13 + 0.85

a:P < 0.01

tumour patients studied . Moreover serum albumin did not differ significantly in the three groups of breast tumour patients as compared with that of healthy subjects .

In vitro studies of 1251-hPRL binding revealed strongly the involvement of calcium in the binding process . Determination of calcium in tumour tissue extracts of breast tumour patients demonstrated no differences between benign and malignant tumours .On the other hand serum calcium of these patients were increased significantly as compared with that of normal subjects •These results may indicate that calcium either, acts in vivo indirectly to stimulate the binding of prolactin with its receptors o it is not involved in this process .This observation needs further clarification .

Several evidences have been suggested to Identify the etiologies of hypecalcemia in cancer patients whom metastasis were not involved in any site (131). Studies have shown that the factors responsible for most cases of this abnormality is a family of polypeptides , related to transforming growth factor (TGF). This factor is able to enhance local PgE_2 production by bone (132). The latter causes , bone resorption and thus, hypercalcemia may be developed (131) .

Malignancy is one of the most frequent etiologies of hypercalcemia folloWing only hyperparathyroidism (133).Actually it is a common complication in cancer patients, evaluation of its incidence varying from 10 to 20% (134) Hypercalcemia is not responsible for gastrointestinal ,renal and neurologic symptoms, but it is also potenially lethal .Thus diagnosis and treatment of this abnormality must be considered .

3.15:1nvestigation of Serum Lipid Profile in Breast Cancer Patients:-

several evidences have been accumilated to suggest the risk of high dietary lipid intake and hence, the elevated lipid levels in the development of breast cancer (135) . Promotion of mammary tumourigenesis has been considered to be mediated by an interactive effect of dietary fat and prolactin (136). In view of this consideration and for the clarification of prolactin receptors role in tumourgenesis of the breast , serum lipid profile was investigated through the determination of serum cholesterol and triglycerides in the three groups of breast tumour patients described in section 3 - 1 There were no significant differences in body weight and age between cancer and non cancer indMduals of the three groups investigated . Table 3-6 shows that serum trIglycerIdes was increased significantly (P< 0.001) in the two groups of the cancerous patients as compared with that of control groups_,whereas serum cholesterol was increased also significantly somewhat less than that of trIglycerides (P< 0.01) . Serum lipid profile was the same in patients with benign breast tumours as in the control group . There Is no correlation found between serum lipid profile and prolactin levels, while there is a significant positive correlation (Fig.3-14) found between prolactin receptors and serum lipid profile In the two groups of breast cancer patients (r ＝ 0.78 , P< 0.01, r ＝ 0.68 , P < 0.01 for triglycerides in groups I and II respectively; r ＝ 0.65 , P< 0.01 and r ＝ 0.68, P < 0.01 for cholesterol in groups I and II respectively) .

The finding of higher lipid levels in breast cancer patients Is in

Table 3 - 6 : Serum Cholesterol , and Triglycerides in Breast Tumour Patients of the Three Groups Studied , Compared to the Healthy Subjects .

	No. of Cases	Age (year)	S.Cholesterol (mg/100 ml)	S.Triglycerides (mg/100 ml)
Group 1	25	$33 \pm .4\ 1$	251 ± 11	$169\ 3 + 4.6$ [a]
Control	21			116.3 ± 8.7
Group 2	24	$54 + 5.2$	b	a
Control	21	$53 + 4.7$	$234 + 7.1$	$131\ 2 \pm 4.2$
Group 3	2	45 ± 7.5	$231\ :;\ 9.5$	$1305 + 5.3$
Control	2		$223\ :t\ 8.7$	

a: P < 0.001

b: p < 0.01

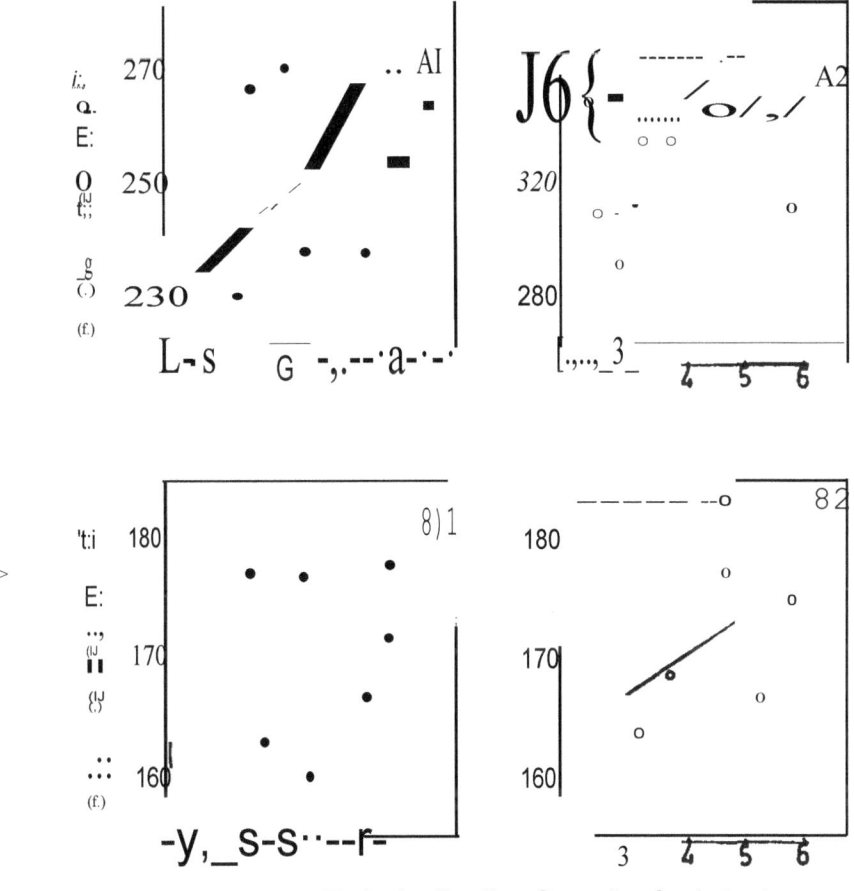

Prolactm Bmding Capacity Cfmole /mg)

Fig 14 : A1,2: Correlation Between Serum Cholesterol and Prolactin Receptors Concentration In Premenopausal and Po tmenop.1u.sal Flrcusf Cancer Patients Respectively.B1,2: A.s in A1,2 but for Seru:n Trlglyct:ridP. 4/. Details arre Explained in the Text

agreement with the findings of Bani et at. {136) who had reported elevated concentration of plasma total lipids , phospholipids ,fatty acids and cholesterol in patients with this disease .However Feldman and Carter (137) reported a decrease in the levels of cholesterol in breast cancer patients _ The conflicting of the results may be ascribed to the type of patients investigated by other authors, since they carried out their study on patients following surgical removal of tumours . Previous studies have supported changes in plasms lipid patterns in many patients following surgery , as part of the neuroendocrine and metabolic response to trauma (138) .In the present study , serum lipid mesurments were performed preoperatively . Any differences found can not be attributed to the effects of surgery,. anaesthesia.... drugs or recovery.The results obtained therefore clearly reflect differences associated with the disease , itself and strongly suggested the implication of the role of high level of lipids in the tumourigenesis of the breast.

The involvement of lipids in the tumourigenesis of breast has been investigated to demonstrate severAl observations. In postmenopausal women, overweight has been linked to a greater risk of breast cancer , and there is a poorer prognosis In pre- and· postmenopausal overweight pateints (139) . There Is a higher Incidence of breast cancer In women married to man who showed higher fat intakes (140) .In addition to the role of dietary effect ,several abnormalities in lipid metabolism h ve been found to associate with breast cancer (141) .

The results obtained for serum prolactin support the significantly higher level of this hormone in breast cancer patients, although there Is no correlation found between serum prolactin and lipid profile. However , the positive correlation found between prolactin receptors and serum lipid profile considerably suggestes the mediation of prolactin receptors in the tumourigenic effect of lipids on the breast. Several observations have been shown to elicit that disturbances in lipid metabolism in breast cancer patients may be related to the elevation of prolactin (136) . Prolactin is reported to increase glucose uptake, fatty acid synthesis in adipose tissue (142) and lipogenesis In liver(143). It is also reported to cause lipid hydrolysis in adipose tissue (142) and raises fatty acid levels in children (144) .However from these observations we can conclude that dietary induced changes in prolactin secretion may provid the mechanism by which plasma lipid alterations occur in this disease. Prolactin receptors appeared to be implicated strongly within this mechanism but the reason , by which prolactin receptors not the hormone itself correlated with serum lipid profile remains unclear.

Conclusions

1 Tt·P. result:·, .)iltamed on the levf:>l of ser11m prolat:ttn 1n ttte cancerous (premenop<1tJSal and postmenopausal) anc1 be 1ign groups of patients studied sugge t that prolactin may be useful in the diagnosiS and management of breast cancer .

2. The binding of prolactin with Its rece!1tors in breast tumours was depended on the conditions of incubation (protein concentration, time, temperature . pH and salts). These results emphasize the presence of true receptors of prolactin in tumour mammary tissues .-

3. The apperance of prolactin receptors in about 40% of breast cahcer tumours and the low availability in the benign tumours (8%) moreover the signifiGant increase of these receptors in the cancerous patients . indicate that the evaluation or prolactin receptor status 11ay provide useful diagnostic informations for breast cancer .

4. The increased levels of serum cholesterol and triglycerides in the two groups of cancerous women and the positive correlation between prolactin receptors and serum lipid profile suggest that interactive effects of lipid and prolactin receptors may be responsible for malignancy of the breast at least in some women . However the assessment of serum lipid profile may be of clinical significance in women with high risk factors or breast cancer .

References

1. Disaia PJ & Creasm<'lll 'NT. (1989) Breast Dim:ases& Colorectal Cancer In: Clinical Gynecologi• Oncology Third ed . .The C.V Mo:. y r.t>mpany Toronto

2. Chetty U. (19"19) Br J Surg   f\7 · 7R9

3. Giuliano AE. (1984) Breast . In Current Medical Diagnosis & TreatmP.nt. Eds . Krupp M,L\ & Chntton . MJ. Middle East Edition . Lebmom . P .429

4. Cutler SJ .(1974) Semi Oncol 1·91

5. Anderson DE. & Badzior.h MD .("1985) Cancer,.56 ·383

6. Gilbertsen VA . (1975) Semin Oneal . 1 : 87

7. Leis, HP (1975) J Reprod Mm1 14 : 231

8. Devitt JE. (1981) Surg Gyneco: Obstet. .152: 4'37

9. Van Oongen JA (1989) Acta Oricol . 78 : 123-3-1

10. Vcheir H (1984) AM J Obstet Gynecol 148: 1?7

11. Tabor L (1985) lancet 13 : 829

12. Cogas J & Shalkeas (1975) Su;gery 73: 339

- 13. Shapiro S , Strax P & Venet L .(1971) JAMA 215 : 177

14. Donegan WL., Spratt JS. (1979) Cancer of the Breast, 2nd   d. Saundrs.

15. Early Breast Cancer Trialists Ct)!laborative Group (1992) La'1cet 339 :1-15 ₇ 71 - 85'

; 16. Sipgletr;n 'J\fW & McCarty KS (1987) Gyneco! Onco\ 26 :?71

":.7. Gelb r Rn & GoldhiP>r:.h A (19 6) J Clin Oncor 4: 169Fi

18. Helman S (1983) Controversies in Bro:-:? t Carr-f!r . Conference Sponsored by the MD Aderson Hospital and Tumour   Institute Chicago, Year   Book ₇ Medical Publisheis , Inc .

, 19. Harris JR , Leven MB, & Hellman S. (1978) Semin Oncol 5:403

.. 20. Butta A , Maclennan K, Flanders KC, &et al (1992) Cancer Res 52:4261

21. Powles T. (1988) Lancet ii:345

22. Slevin ML, Stubbs L, Plant HJ & et al. (1990) Br Med J 300:1458.

23. Banadonna G . (1985) Breast Cancer Treat 5 : 95

24. Carbone PP. (1975) AM J Clin Pathol 64: 774

25. Saunders CM . (1993) BJHM 50: 588

26. Haagensen CD, Bodian C & Haangensen DE (1981) Breast Carcinoma Risk and Detection, Saunders .

27. Chetty U .(1979) Br J Surg 67 : 789

28. Cole P, Elwood JM, & Kaplan SO. (1978) AM J Epidemiol 108: 112

29. Ariel IM .(1973) AM J Obstet Gynecol 117:453

30. Oluwole SF, Freeman HP. (1979) AM J Surg. 37: 786

31. Rao BR .(1981) Cancer 47:2016

32. Ricciardi I. & Ianniruberto A. (1979) OHtet Gynecol 54 : 80

33. Kramerr WM, & Rush BF .(1973) Cancer 31 :130

34. Devitt JE .(1972) Surg Gynecol Obstet 134: 803

35. Leis HP. (1970) Diagnosis & Treatment of Breast Lesions . New York Medical Examination Publishing Co. , Inc .

36. Baker HW, &Snedecor PA. (1979) AM J Surgery 45 : 727

37. Kline TS ,Joshi LP & Neal HS .(1979) Cancer 44 : 1458

38. Marshall J, Graham S, & Swanson M. (1982) AM J Public Health 72: 610

39. Gonzalez ER (1981) JAMA . 244 · 1077

40. Nezhat C. (1980) AM Obstet Gynecol 137 : 604 •

41. BUchert- Toft, M .(1979) Br Med J 1 : 237.

-+-42. Yamamoto K (1985) Ann Rev Genet 19;209- 252.

43. Schwartz M.K (1976) Cancer 37 : 542 .

44. Schwartz M.K (1975) Cancer 36: 2334 •

45. Reddik K and Holland JF (1976) Proc. Nall Acad Sci USA
     73: 2308.

46. Henderson M & Ke::sel 0 (1977) Cancer 39 : 1129 •

47. Bhattachara M, Chatterjee SK & Barlow JJ (1976) Cancer

                                                    Re

     s 36: 2096.

48. Bauer CH, Reutter WG, Erhart KP & et al . (1978) Science 201 :1232 •

49. Wood CH., Varela V ,Palmquist M ET al .(1977) J . Surg Oncol 5: 251.

50. Wilson RG , Buchan R , Roberts MM & et al . (1974) Cancer 33 : 1325 .

51. Secreto G, Recchione C, Cavalieri A & et al (1983) Br J

                                                    Cancer

     47: 269.

52. Barbara HM , Holdaway IM , Mullins PR & et al.(1983)cancer Research
     43: 2985.

—· 53. Shapiro CM , Schifeling O, Bitran JO & et al .(1982) J Surg Oncol.
     19:119.

54. Li CH (1977) Hormonal Proteins & Peptides : Growth Hormone and
     Related Proteins Vol . IV . Academic Press .

.55. Howannitz JH (1984) Evaluation of Endocrine Function . In. Henry JB
     (ed.): Clinical Diagnosis & Management by Laboratory Methods. W.B
     Saunders, Philadelphia . P.299 .

:1- 56. Bennet BO & \Neils OJ (1985) Clinical Chemistry;Principles Procedures & Correlation. Bishop ed. P.307.

, 57. Abraham GE (1977) Hand Book of Radioimmunoassay,2nd ed. Marcel Dekker , Inc . New york P. 179.

58. Nagasawa H . & Sakai N & Banjerjee MR . (1979) Life Sci. 24 : 193 .

59. Ho KY, Evans WS & Thorner Mo (1985) Clin Endocrinol Metab.14: 1.

60. Nicol Cs & Byrant GD (1972). Physiological & Immunological Properties of   Prolactin . In : Prolactin and Carcinogenesis . Alpha Omega Cardiff P.72

61. Neville MC & Neifert MR (1983) Lactation Physiology , Nutrition & Breast Feeding . Plenum Press New York. P.219

.i.- 62. Zilva JF.., Pannall PR & Mayne PD (1988). Clinical Chemistry In Diagnosis & Treatment

63. Cowie AT, Forsyth IA & Hart IC. (1980) Hormonal Control of lactation Springer Verlag, Berlin P.801

64. Shiu RPC & Friesen HG (1980) Ann Rev Physiol. 42 : 83

65. Mcmurty JP & malvern PV (1974) J Endocrinol 61 : 211

66. Bartke A. (1980) Fed. Proc 39 : 2577

67. Bex FJ, Bartke A, Goldman BD & et al.(1978) Endocrinology 103: 2069

68. Deshpande N. (1975) J . Stroid Biochm 6 : 735

69. Sheth NA, Ranadive KJ, Suraiya JN & et al .(1975) Br J Cancer 32:160

.70. Trichopoulos D, YenS, Brown J & et aL (_19) Cancer 53: 187

71. Nagia R, Kataoka MI Kobay S, & et al. (1979) Cancer Res 39: 1835

72. Rose DP & Pritt BJ. (1981) Cancer 48 : 2687

730 Bani IA, Williams CM, Boulter PS & et al 0(1986) Br J Cancer 54 : 439

74. Rolandi E , Berrence T, Masturzo P & et al 0 (1975) Lancet. ii: 845

750 Cole EN , Sellwood RA, england PC & et al 0 (1977) Eur J Cancer 13:597

760 Krishnamoorthy G, Govlndarajulu P & Ramal in ſ Vn (1989)Neoplasma 36: 221

770 Jones MK, Ramsay 10, Collins WP & et al 0 (1977) Eur J Cancer 13: 1109

780 Mafadyen IJ, Forrest·APM, Prescott RJ & et al . (1976) Lancet ,i :1100

790 Copper JR, Bloon FE, Roth RH (1978) Receptors 0 In,The Biochemical Basis of NeuropharmacologyoOxford University Press . Inc. Po60

800 Granner DK (1988) Hormone Action 0 in : Harper's Biochemistry 0 eds . Murray ROK et ai.Tewenty- first edt. Middle East Ed. P.472

810 Gilman A.(1984) Cell 36 : 577

820 Yamamoto Ko (1985) Ann Rev Genet . 19 : 209

830 Weinberg R 0 (1985) Science 230 : 770

840 Borkowski A , Body JJ & Lenclercq G (1988) Eur J Cancer Clin Oncol 0 24 : 509

850 Osbme Cko (1985) Sem Oncol 0 12: 317

860 Croton R, Cooke T, Holt S, & et al 0 (1981) Bro Med Jo 283: 1289.

870 De Sombre ER, Greene GL, Jensen EV,(1980) Estrogen Receptors & the Hormone Dependence of Breast Cancer In:Breast Cancer :New Concepts in Etiology & Control 0 New York, Raven Press

880 Banifacino JS , Dufau Ml 0 (1984) Prolactin Receptors in Ovary In Saxena BB (edo) Hormone Receptors in Growth & Reproduction . Raven Prees, New York

89. AL- Khayat TH. (1991) Ph.D Thesis . University of Baghdad Colleg e of Science

90. Cohen R & Ashkenazi A, Elbery G & et al. (1987) J Recep Res 7:921

91. Djiane J , Houndebine L,Kelly PA (1985) Proc. Natl. Acad Sci 78: 7445

92. Kelly PA, Djiane J & Leblanc G .(1983) Proc. Soc EXP Bioi Med 172 : 219 .

93. Shiu RPC & Friesen HG .(1974) Biochem. J. 140: 310

94. Aragona C & Bohnet HG. (1975) Endocrinology 97 : 677

95. Djiane J , Clauser H & Kelly PA . (1979) Biochem .Biophys Res Commun . 90 : 1371

96. Bradly CJ, KLEDZIK GS & Meites J (1976) Cancer Res. 36 : 319

97. Welsch CW. (1985) Cancer Res 45: 3415

98. Salih H , Flaxh , Bronder W . & et al . (1972) Lancet 2 : 1103

99. Barrett A, Desouza I, Morgan L & et al (1975) Lancet 1 : 1347

100. Partige RK & Hahnel R .(1979) Cancer 43: 643

101. Holdaway IM & Friesen HG (1977) Cancer Res. 37: 1946

102. Stagener Jl, Jochimsen PR 7 Sherman BM.(1977) Clin Res 25: 302A

103. Balrlaux ML, Costells S, Vokaer A & et al. (1984) Prolactin & Prolactin ecep ors in Human Breast Dieases In : Progress in Cancer Research & Therapy Eds . Bresciani F & et al. Vol.31 New York Raven Press

104. Bonneterre .J , Peyrat JP, Beuscart R & et al (1987) Cancer Res 47 : 4724

105. lowry OH, Rosebrough NJ, Farr AI & et af. (1951) J Bioi Chern. 193: 265

106. Morris BJ (1976) Clinica   Chimica Acta. 73 : 213

107. Haro L & Talamantes F. (1985) Mol Cell Endocrinol 43: 199.

108. Scatchard G (1949) Ann N . Y . Acad Sci. 51 :660

109. Varty H. (1980) Practical Clinical Biochemistry ,Fourth ed. William Heinemann. Medical Books. ITD   New Yourk

.110. Varly H.(1988) Varly 's Practical Clinical Biochemistry. Gounlook M ed. 5th eds. Medical Books LTD. New Yourk.

111. Fossati P & Prencipe L (1982) Clin Chern 28 : 2077

112. Secreta G, Recchione C, Gavalieri A & et al . (1983) Br J Cancer 47: 269

113. Birkinshaw M & Falconer IR. (1972) J Endoc'rinol 55: 323

114. Lineweaver H & Burk D. (1934) J Amer Chern. Soc. 56: 658.

115. AL- Mahdawi FM (1992). Ph.D. Thesis ,University of Baghdad. College of Science

116. Melander W. & Hovarth C. (1977) Arch . Blochem. Biophys . 183 : 200

117. Ahamd BM. (1992) Ph.D Thesis.Universlty of Baghdad. College of Science.

118. Waelbroeck M , Obberghen EV & De Meytes P.(1979) J . B.C 245: 7736

"./ 11 ; Jarr:ott B. & Adam.s A. (1985) Biochem. Biophys Res Cornm 128:816

120. Malarkey WB, Kennedy M. Albred LE, & et al. (1983) J Cfin Endocrinol Metab. 56: 673 .

121. Holtkamp W, Wonder HE, Von Heyden D, & et al. (1983) J Ster Biochem (1983) 19 : Suppl. : 143S

122. Manni R, Rainieri J, Arafah BM & et al (1982) Cancer Res. 42 : 3492

123. Peyrat JP, Dewailly 0, Djiane J & et al. (1981) Breast Cancer Res. Treat. 1 : 369 .

124. Bonneterre J, Peyrat JP Van de Walle B, & et al. (1982) Eur J Cancer Clin Oncol -: 18 : 1157 . .

125. Turcot - Lemay I& Kell PA. (1980) Cancer Res. 40 : 3232

126. Djiane J , Durand Ph & Kelly PA (1977) Endocrinology. 100 : 1348

127. European Breast Cancer Group. (1972) Eur. J . Cancer 8 : 155

128. Arafah BM, Roe J, Manni A & et al. (1982) Endocrinology 117 : 584

129. Shafie S & Brooles DI (1977) Cancer Res. 37 : 794

130. Edery M, Goussayd 0, Vives C & et at. (1983). Biomedicine.33: 265

131. Valentin - Opran A & Meunier PJ . (1984) Hypercalcemia of Malignancy : Pathogenesis & Treatment . Im, Progress in Cancer Research & Therapy. Eds. Brescianl F. & et al. V.31 New York Raven Press.

132. Tashjian AH, Voelked EF, Goldhaber P.& et al. (1978) Biochem. Biophys Res. Commun. 85: 966

133. Mundy GR, Cove DH, & Fisken R. (1980) Lancet i : 1317

134. Sporn. MB, Todaro GJ (1980) N.Eng J Med. 303: 878

135. CHan EN & Dao TE. (1981) Cancer Res. 41 : 164

136. Bani IA, Williams CM, Boulter PS & et al. (1986) Br J Cancer 54:439

137. Fildman EB & Carter AC.(1971) J Clin Endocrinol Metab. 33 : 8

138. McNamara JJ, Molot M, Durran EL & et al. (1972) J Thoracic Cardia Surgery. 63: 968

139. Greenberg ER, Vessey MP, Mcpherson K, & et al (1985) Br J Cancer . 51 :691

140. Nomura A, Henderson BE & Lee J (1978) AM. J . Clin Nut 31 :2020

141. Wood CB, Habib NA, Thompson A & et al. (1985) Br Med J 291 : 163

142. Hamid Ma, Rubenstein D, Fergusan KA & et al. (1967) Biochem. Biophys. Acta - 100 : 179

143. Macleod RM, Bass MB, Hwang Sc & et al. (1968) Endocrinology 82: 253

144. Elsiar J & Denine R, (1970) Rev Eur Etud Clin Bioi. 15 : 899